D

Environmental Pollution Studies

GERRY BEST

former Head of Chemistry
Scottish Environment Protection Agency

LIVERPOOL UNIVERSITY PRESS

First published 1999 by
LIVERPOOL UNIVERSITY PRESS
Liverpool L69 3BX

British Library Cataloguing-in-Publication Data
A British Library CIP record is available

ISBN 0-85323-923-1 (paperback)

Typeset in 10.5/14pt Meridien by
XL Publishing Services, Lurley, Tiverton
Printed by Bell and Bain Limited, Glasgow

Environmental Pollution Studies

Cover Illustrations
1. Activated sludge treatment works
2. Algal scum
3. Collecting silage from fields
4. Sea cages of a salmon farm
5. A year's rubbish from a UK family
6. A compactor working on a refuse tip
7. River Girvan downstream of the mine-water discharge
8. Reed beds for purifying ferruginous mine water
9. High stack for dispersing airborne pollutants
10. Skeletal remains of diatoms in sediment *[courtesy Dr Roger Flower, Environmental Change Research Centre, University College, London]*
11. Road traffic on the M8
12. Lichen on mature trees

TO WOJI
WITH LOVE

Contents

Acknowledgements

I would like to thank a number of people who have helped me prepare this book. My colleagues at SEPA, particularly Calum McPhail who gave helpful comments and suggestions after reading the draft, and Liz Allen, SEPA West's Librarian who patiently looked out various references and relevant books.

Particular thanks are due to Shelda Pirie, Head of Chemistry at Hamilton College, who spent a lot of time evaluating and testing the experiments as well as offering useful comments on the script.

Anne Wilson quickly came to my aid in preparing the many illustrations in such a clear manner.

Throughout it all, my wife, Alycia, has given me much needed encouragement and support for which I'm very grateful.

Gerry Best
July 1998

Foreword

Our world, the Earth, is a relatively recent creation in terms of age of the universe. This planet of ours was formed many, many millions of years ago, but life has been present on the Earth for about 4,600 million years. The human species, 'Homo Habilis', is a fairly recent phenomenon having only inhabited the Earth for a mere 3 million years! If we put those 3 million years into the well-known context of a 24-hour clock, then ancient man slowly developed and evolved along with dinosaurs and pterodactyls and the like, for most of the day and evening, the dinosaurs became extinct at ten to twelve midnight, the civilizations of ancient Greece and Egypt emerged at just three minutes to twelve, Christ was born at one minute to midnight, and the last hundred years have passed by in the dying three seconds of our 24-hour day of human existence on our tiny Earth.

And yet, in those last dying three seconds of our day, we have managed to damage our environment to a greater extent than in all the previous 23 hours, 59 minutes and 57 seconds, of our existence.

During most of our time on Earth we have kept it fairly clean and did not begin to damage the environment until very recently. Indeed until the middle of the nineteenth century there was relatively little pollution. But, as you will find out in this book, there are many more of us today. There are more people alive now than the sum total of all those who have lived, and died, since people first appeared!

There are so many of us, and we want so much, our cars, our buses, our comfortable and convenient ways of life, that the Earth – our world – is beginning to become dirty, soiled and polluted.

It is time to do something about this pollution of our environment and this book is a very valuable contribution to this problem facing our communities today. It explains current problems and suggests answers and provides a possible way forward in important areas of concern. In reading this book you will learn something of the 'down-to-earth' side of pollution, the areas that really matter, about sewage collection and disposal, about farm pollution and eutrophication, about water pollution, about pollution of the air and our atmosphere and so on. Environmental education is, belatedly, becoming an essential part of every young person's educational process.

There is a very simple, old Native American saying, which has been often quoted, and indeed copied, but which is in my opinion still the best expression of the key reason why we must stop polluting our planet. It states:

'We do not inherit the earth from our ancestors, we borrow it from our children.'

Professor W.A. Turmeau
Chairman, Scottish Environment Protection Agency

1. Introduction

Water pollution, global warming, poor air quality, acid rain, holes in the ozone layer, etc. – these are issues that are featured regularly in our newspapers, news reports and TV programmes. Ever since we received the first pictures of the earth from outer space, we have become much more aware of how small our world is in the cosmos and how vulnerable it is to destruction by human activities. Concern about the state of our environment is now one of the main issues in people's minds, even higher up the list than war, unemployment and health.

National governments have responded to these concerns and most of them now have specialist Environment Ministries to ensure that laws are enacted, not only to protect the environment of their countries, but also to contribute to world-wide efforts at reducing the awful damage that the human race is doing to the natural world.

The pressures to do so are enormous because, not only are our numbers increasing at a great rate but we all look for a better standard of living. The figures in Table 1 illustrate the scale of the problem now, and for the future.

Table 1. World population, 1930–2050 (millions)

Year	Population
1930	2,070
1960	3,020
mid-1990s	5,300
2000	5,720
2050 (projected)	9,830

Each of these people is looking for the basic necessities of life – food, a home and clothes to protect them from the elements – but many hope for much more. In the UK for example, each family wants a TV, radio, a range of furniture, cooker, washing machine, 'fridge, car, computer, holidays and so on. Each week, millions of pounds are gambled on the National Lottery in the hope that it will provide a short-cut to get these things (and lots more!). Each of these items though makes some impact on the environment, whether it be in their manufacture or their use, and the more we have, the greater our effect on the natural world. It has been said that an increase in the population of the developing countries (Africa, India, etc.) is a problem for the world but an increase in the number of people in the USA and Western Europe is a disaster because of the greater demands each one of them makes on the earth's resources.

There hasn't always been such concern about environmental pollution.[1] At the height of the Industrial Revolution in the middle of the nineteenth century, there were serious pollution problems in the air, water and on land. At that time, the black smoke belching out of factory chimneys was seen as a sign of prosperity; likewise with the discharges of waste entering rivers through sewer pipes and rubbish dumped into them. The priority was to produce the goods that people wanted using the new technology of steel manufacture, wool and cotton mills, tanneries, dye works, etc. 'Where there's muck there's money' was a common expression. The poor quality of air, working conditions, and water supplies took its toll on the residents in the towns and cities, and the life expectancy was low.

The condition of the UK's rivers deteriorated rapidly in the period 1800–50 as towns grew quickly with the influx of workers from the country. Sewage treatment for human and industrial waste was minimal; most was piped directly into rivers from newly constructed sewers. One of the most polluted rivers in Britain was the Thames: conditions were so bad in the hot summers of 1858 and 1859 that the debates in the Houses of Parliament were suspended because of the awful smell coming from the river outside the building.[1] Disease was rampant, particularly cholera and typhoid. Indeed, it was the pioneering work by John Snow and William Farr in London in the 1850s that established that cholera was a water-borne disease. In 1861, Queen Victoria's husband Prince Albert died from typhoid after drinking contaminated water. His death, and the terrible state of rivers and water supplies, spurred Parliament to improve the quality of the environment.

A Royal Commission was set up in 1865 to 'enquire into the best means of preventing pollution of rivers'. It produced six reports in all and, in its fourth one in 1872, it recommended various standards of cleanliness that effluents should achieve as well as legislation to enforce them. These eventually were incorporated into the Rivers Pollution Act of 1876. Unfortunately, the legislation was ineffective for two main reasons. Firstly, the enforcers of the standards were the local authorities – who just happened to be the principal polluters! Secondly, there was an 'escape' clause for industry in that, if the installation of pollution control equipment was too costly for the company, then it could be exempt from prosecution. Despite this, the Act remained in force for 75 years – and rivers did not get any cleaner.

The real impetus to improvements came in the 1950s and 1960s as a result of modifying the legislation and setting up independent pollution prevention authorities. In England and Wales these were the River Authorities and in Scotland the River Purification Boards. The new authorities set about enforcing the new laws and prosecuting those who caused severe pollution. Gradually water quality has improved to the extent that fish have started to return the rivers. In 1983, coincidentally, the first salmon for over a hundred years were recorded in both the rivers Clyde and Thames.

Pollution is a term that has been used quite freely for many years without a clear definition. Now though, it is generally accepted that pollution of water can be defined as 'the discharge by man, directly or indirectly, of substances or energy into the aquatic environment, the results of which are such as to cause hazards to human health, harm to living organisms and to aquatic ecosystems, damage to amenities or interference with other legitimate uses of water'. This definition can be modified to apply to other sectors of the environment such as the land or the air.

Recently, the pollution prevention management of the UK's environment has undergone another fundamental change. Since April 1996, the protection of air, water and land against pollution has been carried out by the Environment Agency (EA) in England and Wales, and the Scottish Environment Protection Agency (SEPA) in Scotland. In Northern Ireland, this work is undertaken by the Environment and Heritage Department of the Department of the Environment, Northern Ireland. (The addresses of these agencies are given at the back of the book.)

One important new role for the Agencies is to promote 'sustainable development'. This idea developed from the United Nations Conference on Environment and Development held in Rio de Janeiro, Brazil, in 1992. Usually referred to as the Rio Earth Summit, this conference was attended by representatives of 178 governments. They pledged to reduce the over-exploitation of the world's resources, to tackle the issues of global warming and to ensure the safety of the environment for future generations. This last aim is the essence of sustainable development: that any developments that will meet the needs of today will not compromise the needs of future generations.[2]

The issues associated with sustainable development are now being incorporated into new laws. For example, the Labour Government which was elected in May 1997 pledged to reduce the numbers of cars entering UK cities because of the pollution they cause and the fossil fuels they use up.

If we look back at the history of environmental legislation we see a distinct shift in emphasis:

Time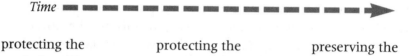

| protecting the people from the environment | protecting the environment from people's activities | preserving the environment for future generations |

We are only just embarking on the last stage; time will tell whether we are successful. The problems are immense, not least because of our selfishness. We may all think that it's a great idea to make the world a better place, but when that has to be translated into making personal sacrifices such as giving up cars, cutting back on flying abroad, paying more for fossil fuels, reducing our use of packaging, including bottles and cans, then it gets more difficult. Individually, however, we can all do something. It is well summarized in the expression: 'think globally, act locally'.

In this book, many of the issues that are threatening our environment are discussed. Additionally the last chapter suggests some projects that can be carried out with simple equipment to measure pollutants and their effects.

Notes

1. Leslie Wood, *The Restoration of the Tidal Thames*, Adam Hilger, Bristol, 1982.
2. *Sustainable Development, the UK Strategy*, HMSO, 1994; Richard Bailey, *An Introduction to Sustainable Development*, Chartered Institution of Water and Environmental Management, London, 1997; and Jennifer Elliot, *An Introduction to Sustainable Development. The Developing World*, Routledge, London, second edition, 1999.

2. Sewage collection and treatment

In our day-to-day living of getting up, having breakfast, going to school, eating lunch and tea, washing ourselves, clothes and dishes, etc., we each use 140 litres of water a day, on average. The greatest contribution to this usage (30–40 per cent) comes from the flushing of the toilet to get rid of our bodies' waste. It seems odd that water is purified to a stage where it's fit to drink and then most of this good quality water is flushed down the loo! In the UK, the modern WC cistern holds 9 litres (2 gallons) but there are plenty of houses with cisterns of 13.6 litres (3 gallons) capacity. In recent years there have been concerns about water shortages in summer because of drought. One way of reducing water shortages is to cut back on our water usage, and this could be partly achieved by reducing the capacity of the WC cistern. In 1996, the House of Commons Environment Committee recommended that people should be encouraged to replace their cisterns with ones of 6 litres capacity. There are also devices now available that can be plumbed into houses to collect 'grey water' (water that comes from washing machines and sinks). This is partly purified and pumped to a storage tank in the roof from where it can be used to flush toilets.

Apart from the WC, the average water consumption, in litres, for different uses in the home is:

Automatic washing machines	100
Dishwashers	50
Bath	80
Shower	30

In addition to the water that is used by ourselves in our homes, the drains also receive the water used by industry and, on average, this amounts to the equivalent of about 300 litres per head per day. Whatever product is being made in a factory, there's almost certainly wastewater being produced which needs to be purified.

Sewerage systems

In a typical town, the sewage from homes and the wastewater from industry enter a sewer – an underground pipe which joins up with others to take the polluted water to a sewage treatment works (STW). The network of sewage pipes is called the sewerage system. In Britain there are two types of sewerage systems: 'combined' and 'separate'.

Combined sewerage systems

The majority of old towns and cities have a combined sewerage system. This means that the rainwater run-off from roads, pavements, roofs, car parks and other hard surfaces is collected in the same sewer as the foul drainage from homes and factories and is then piped to the STW.

In wet weather, the sewage is diluted by rainwater and the amount of sewage entering the STW is increased considerably. If the rain is prolonged, the volume of sewage entering the works exceeds its capacity. Some of it can be diverted into special holding tanks called storm tanks where it is stored until the storm is over. It can then be slowly emptied into the treatment plant to be purified. If, however, the storm tanks become full, then the overflow from them passes into the river. By that stage the river itself should be full of rainwater so the untreated sewage is well diluted.

The problem with the combined sewerage system is that the STW has to be large enough to cope with the high rainfall in storms and these don't happen very often. So the storm tanks are empty most of the time, and it costs a great deal of money to build them. Also, the nature of sewage arriving at the STW varies according to the weather and this is not good for the purification process (which is described on p.8). For these reasons, separate sewerage systems are often found in new towns and developments in Britain. In Canada and the USA, separate sewerage systems have to be provided.

Separate sewerage systems

In separate sewerage systems, the foul water from homes and factories, etc. is collected in one sewer and piped to the STW for treatment. The surface-water drainage is collected in a separate sewer and piped to the nearest river or stream. Theoretically the surface water should be relatively free from pollution, but in reality it contains oil and silt from roads and it can be polluted by spillages in factory yards. In winter months, roads are often treated with salt and grit to remove ice and snow. When these melt and the roads are washed by rain, the receiving rivers have increased concentrations of sodium and chloride from the salt and are turbid because of the grit.

Whatever sewerage system is installed in a town, the most polluted water is directed to a treatment works to be purified before being released into the environment. The amount of treatment depends on the location: in the UK, all inland towns have STWs that give full biological treatment to the sewage, whilst for coastal communities, the treatment is minimal, usually only filtering it through a wire-mesh screen, before the sewage is released into the sea.

Sewage treatment

Prior to the development of the modern sewage purification process, most sewage was disposed of onto land usually situated on the edge of towns and villages. The land was deeply ploughed and the sewage was allowed to flow along the furrows and soak into the soil. Sewage contains valuable nutrients such as phosphorus and nitrogen so the soil became very fertile. Vegetables were grown on the ridges of the furrows and this gave rise to the term 'sewage farms'. This method had the merit of putting the sewage to good use, but because of the rising cost of land, the demand for more land for housing and the problems of unpleasant smells, sewage farms were superseded by other methods of treatment.

The main methods of sewage treatment used today depend on the breakdown or oxidation of the organic matter in the sewage by bacteria in the presence of oxygen from the air. In other words, it's a biological process. In fact, the purification of sewage duplicates the natural degradation process in a river, except the process is intensified and speeded up.

When organic matter enters the water in a river naturally, for example the leaves that fall into it in autumn, the bacteria present in the river bed or dispersed in the water gradually break down the leaves into their basic components and these provide food or 'building blocks' for other life in the water.

The bacteria which carry out this degradation process in the presence of oxygen are called aerobes whereas those that break down organic matter in the absence of oxygen are called anaerobes. In the sewage treatment process, in order for the aerobic bacteria to operate efficiently, it is important that as much oxygen as possible comes into contact with the sewage. Modern sewage treatment processes are designed to bring this about. In the percolating filter system, the sewage is spread over the surface of large tanks filled with stones. The sewage becomes a thin film which results in a large surface area for the gaseous diffusion. In the activated sludge process, the sewage is aerated either by pumping in bubbles of air or else by stirring vigorously.

Percolating filter sewage treatment works

For many years, the percolating filter system of sewage treatment was the most commonly used (although now it is steadily being replaced by the activated sludge process). The incoming sewage, whether it be from a combined or separate sewerage system, is first passed through a 'gate' of parallel iron bars (coarse screen) which removes the large objects that find their way into the sewers, such as pieces of wood, bottles or cardboard boxes. The grit and stones from roads etc. are allowed to settle out in a small tank and the sewage then flows into very large tanks known as primary settlement tanks. Here, the heavier particles of organic matter settle to the base of the tank as a thick muddy layer which is known as primary sludge. This sludge is pushed into a central chamber by scrapers attached to a large beam which is slowly rotated round the tank by a motor attached to its end, and then it is siphoned off. About 50 per cent of organic matter in sewage is removed in the primary tanks.

Whilst the sludge accumulates at the base of the tanks, the cleaner surface layers overflow from the primary tanks on to the next stage of treatment where the sewage is purified by micro-organisms. The settled sewage is spread over the surface of further tanks, which are either circular or rectangular, at least two metres deep and full of stones. The stones are about five centimetres in diameter near the surface and slightly larger further down. The sewage is spread by slowly rotating radial arms in circular tanks, or by a suspended beam driven backwards and forwards across rectangular tanks. The settled sewage trickles down through the tank and over the stones. The stones are covered with a slime of many different micro-organisms, mainly bacteria but also protozoa, worms and the larvae of insects. The bacteria feed on the sewage and purify it, whilst the worms and insect larvae live on the bacterial slime, thereby controlling its growth and preventing it from clogging the spaces between the stones.

The water that collects at the base of the percolating filter tank contains the breakdown products of the purified organic matter, dead microbial cells and insect larvae, and the waste products of the larger organisms in the tank. This purified sewage is piped to more settling tanks, so-called final settling tanks, where the debris from the filters is allowed to settle and the overflow of relatively clean water is allowed to pass to the river. The whole process is shown in Figure 1. Thus the final products of the sewage treatment process are the clean effluent that goes to the river, and the sludge which accumulates in the primary and final settling tanks.

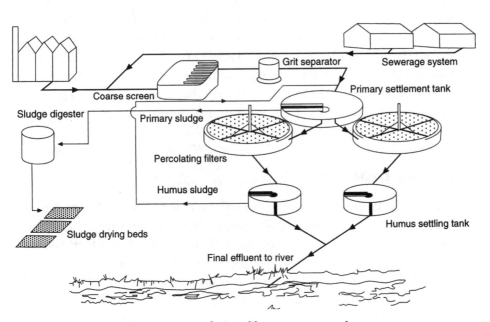

Figure 1. Percolating filter sewage works

Activated sludge sewage treatment works

The activated sludge process is a more modern one and is often favoured because the works occupy less space, it is more controllable and there are fewer problems with flies from hatching insect larvae.[1] As with the percolating filter treatment process, the sewage is initially screened and given primary settlement. Following this it is channelled into large tanks and mixed with activated sludge, which consists mainly of a thick 'soup' of bacteria and protozoa. These organisms break down the organic matter whilst the mixture of settled sewage and activated sludge is circulated about the tank, either with the help of vigorous jets of air bubbles or by being stirred up by rotating stirrers. The oxygen in the air is very important to the process: if the system fails in some way and the bacteria don't get enough oxygen to purify the sewage, the sewage goes 'septic' and smells offensively.

The activated sludge/settled sewage mixture is stirred for about four hours after which time most of the organic matter has been degraded (broken down). It then flows into final settling tanks where the sludge settles to the bottom whilst the clean water left at the top (the supernatant water) flows into the river. The sludge is returned to be mixed with more sewage from the primary settling tanks. However, because the process is a biological one in which the bacterial numbers increase as they feed on the sewage, there is always an excess of activated sludge produced. This excess is mixed with the sludge from the primary tanks and disposed of. The activated sludge process is shown diagrammatically in Figure 2 whilst Cover Illustration 1 gives a general view of one of this type of sewage treatment works.

Tertiary treatment

The sewage treatment processes described above can purify sewage so that at least 90 per cent of the organic matter is biodegraded and the concentration of nutrients is substantially reduced. The effluent can be discharged into most rivers without any adverse effect on the receiving water quality. There are some rivers though that drain into lakes, and others may be shallow and slow flowing. In these cases, the nutrients present in the purified sewage effluent may encourage the growth of water weeds or algae to excessive amounts (see Chapter 3). The excess weeds or algae can be harmful to other aquatic life because they cover

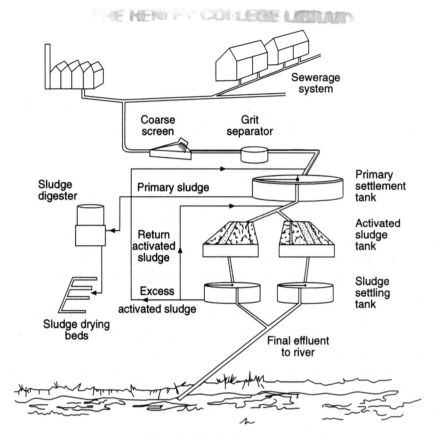

Figure 2. Activated sludge sewage works

the river bed or the water surface and prevent light from penetrating to the deeper water. The nutrient of particular concern is phosphate as the concentration of this is usually the deciding factor in the growth of aquatic plants and algae. These nuisance growths can be prevented by giving the purified sewage an extra treatment stage for nutrient removal. This can be either a physico-chemical process or a biological one.

In the physico-chemical process, a metal salt is added to the sewage. This forms an insoluble compound of phosphorus which settles out in a sedimentation tank. The metal salts added are those of either iron or aluminium. The reaction between aluminium and phosphate can be represented simply by the following equation:

$$Al^{3+} + PO_4^{3-} \rightarrow AlPO_4 \downarrow$$

The most common form of aluminium used is alum which has the formula

$Al_2(SO_4)_3.14H_2O$, and the reaction is as follows:

$$Al_2(SO_4)_3.14H_2O + 2PO_4^{3-} \rightarrow 2Al\ PO_4 \downarrow + 3SO_4^{2-} + 14H_2O$$

Iron salts are commonly used to precipitate phosphate at sewage treatment works and both ferrous (Fe^{2+}) and ferric ions (Fe^{3+}) are used in the form of sulphate or chloride. Their reactions are:

$$3FeCl_2 + 2PO_4^{3-} \rightarrow Fe_3(PO_4)_2 \downarrow + 6Cl^-$$
$$3FeSO_4 + 2PO_4^{3-} \rightarrow Fe_3(PO_4)_2 \downarrow + 3SO4^{2-}$$

There are various biological methods of removing phosphorus from sewage before it is discharged into a river but a particularly attractive one is to construct an artificial 'wetland' or a reed bed. These not only remove the nutrients but also further reduce the amount of organic matter in effluent.

In its simplest form, a wetland comprises a large shallow tank with a very gradual slope extending from a wide distribution system at the inlet end, to a collecting channel at its lowest end. The tank is filled with a bed of gravel into which reeds are planted. The roots of the reeds filter out organic matter and suspended solids whilst the reeds themselves extract nutrients from the effluent before the drainage water overflows into the river.

In some locations the reed beds can be extensive and are constructed by creating large flat areas of ground adjacent to the river or lake. They become havens for wildlife which take advantage of the cover provided by the reeds and exploit the food available in the marshy ground. (In Chapter 7, the use of a reed bed to clean up mine-water drainage is described.)

In some countries, the nutrients in sewage effluent are used to grow other crops. In Scandinavia and Poland, the effluent from sewage works is diverted into large areas planted with willow trees. The trees grow rapidly because of the extra supply of nutrients in the water and they are regularly pruned to provide wood for burning in domestic fires, for power generation or for paper manufacture.

Nowadays, we are much more aware of the loss of nutrients from sewage and how we have been throwing away this valuable resource. The next section describes how sewage sludge is increasingly being recovered and utilized to improve the quality of soil by increasing its nutrient status.

Disposal of sewage sludge

The sludge that accumulates from the sewage treatment processes has the consistency of thin porridge and is over 95 per cent water. Many sewage works receive millions of litres of sewage each day for treatment and this results in the production of thousands of litres of sludge which has to be disposed of. The solid material in it is rich in organic matter which is readily broken down by bacteria and this can give rise to offensive smells. For this reason, sludge is usually pre-treated before it is used for anything.

The treatment of sludge involves breaking down (or stabilizing) the organic matter and removing excess water. In sludge stabilization, the sludge is broken down by anaerobic bacteria, in other words, in the absence of air. It is pumped into large enclosed tanks resembling gas storage tanks (gasometers) and is warmed up to about 35°C to accelerate the process. Methane gas is produced by the anaerobic digestion, and this is collected and burned to produce the heat for warming up the sludge. The stabilization process takes many weeks but when complete the sludge has been converted into a valuable humus.

The next stage is to remove the excess water and this is done either by spreading the sludge out on drying beds or by using vacuum filtration. A sludge drying bed comprises a shallow tank with a thin layer of gravel or crushed stones. The water percolates through the stones whilst the sludge is dried by evaporation over a period of weeks, until it has a water content of about 60 per cent. By this stage it is a manageable solid and can be used for various purposes.

In vacuum filtration, the water in the sludge is removed by filtering the sludge under vacuum through filters made of cloth. Chemical coagulants are added to increase the filtration rate. The sludge adheres to the cloth as a 'cake' which has a moisture content of about 75 per cent water.

For many cities in the UK and elsewhere, including London, Manchester, Glasgow, Edinburgh and New York, the wet sludge from the anaerobic digestors used to be disposed of by putting it into specially designed tankers and dumping it in deep water off the coast. It is now recognized that this is an unsatisfactory method, particularly because the sludge has value as a soil improver and because there was accumulating evidence of environmental damage from sea disposal. In the USA, Congress passed legislation in 1988, the Ocean Dumping Ban Act, that barred the disposal of municipal sewage sludge and industrial waste in the seas from the end of 1991. In Europe, as a result of a European Union Directive, the dumping of sludge at sea was banned in 1998 and alternative disposal techniques found.

In the UK, sludge is now spread on poor quality soils to increase their productivity. For example, it improves the fertility of spoil heaps from coal mining or the poor quality upland soils that are used for commercial forestry. In the Grampian area in Scotland, the sludge is dried to small pellets and bagged and the local farmers and horticulturists use the pellets as a fertilizer. Sludge can also be incinerated because it has a high calorific value and this is the preferred disposal method for some cities such as Edinburgh. The heat released from the burning will be used to dry incoming sludge and to produce power from the steam. Another recent idea is to mix sewage sludge with domestic refuse and then burn it all in special incinerators. The heat produced will be used for heating nearby houses and other buildings. This is called district heating and it is a common method in Scandinavian countries. In a recent UK example, the methane produced by sludge digestion at a sewage works run by Severn Trent Water is collected and burned and the heat is used at the neighbouring Lower Wick swimming baths to keep the water at 30°C – for free!

The chemistry of sewage treatment

If we wish to understand the chemistry of water pollution, we first have to study its origins. In most cases, water pollution is caused by untreated or inadequately treated sewage entering a river with the result that the amount of pollution is too great for the river to absorb. A study of the chemistry of sewage effluent will identify the problem pollutants and the ones to test for by chemical analysis.

Sewage is a complex mixture of organic and inorganic substances, some of these originate from natural materials and some of which are synthetic. Natural materials usually break down much more readily than synthetic ones because the organisms responsible for biodegradation can 'feed' on natural substances but are not so used to synthetic ones.

A good example of the difference between natural and synthetic substances is the bag you put your shopping in at the supermarket. Fifty years ago, when people went to the shops for goods, the items were put into paper bags because plastic wasn't in common usage then. As you know, when paper gets wet, it quickly disintegrates and is broken down into paper fibres which are eventually converted by enzymes into carbohydrates. The modern plastic bag from the supermarket checkout is a synthetic product made from oil and it is non-biodegradable. When it is thrown away it stays as a plastic bag for years. You have only to look at the verges of urban motorways to see how persistent plastic bags are in the environment!

Domestic sewage contains the wastewater from wash-hand basins, kitchen sinks, toilets, washing machines and dishwashers. Apparently, according to a survey in the UK, when we flush our toilets, we dispose of much more than our faeces and urine. The commonest other items to be found are cotton buds, nappy liners, condoms, plastic wrapping, razor blades, sanitary protective items, finger- and toenails and even spiders![2]

Most sewage works also receive the waste from industries in the area, such as food factories, electronics industries, engineering works, chemical manufacturers, textile factories, etc. As it arrives at the works the sewage is 99.9 per cent water, with the remaining 0.1 per cent consisting of dissolved salts, trace metals and a variety of other substances such as soaps, detergents, sugars, food particles, faeces, fats, oil, grease, plastics, clay and sand. All these substances can be classified as organic and inorganic components.

Organic substances
The organic substances in sewage mainly comprise carbohydrates, proteins, fats, soaps and detergents. All of these can be broken down into simpler substances by the micro-organisms in the sewage purification process. Some of these breakdown products are present in the final effluent as it discharges into the river, others are used as the food for the bacterial slime or the activated sludge in the sewage works, whilst the remainder sink to the bottom of the settling tanks to become the sludge.

Carbohydrates are complex molecules containing just the elements carbon, hydrogen and oxygen. They are widely distributed in plants and animals as sugars, cellulose, starch and dextrin. In their journey through the sewage works, they are broken down by micro-organisms into simple sugars, carbon dioxide and water. The carbon dioxide will either vent to the atmosphere or form carbonates and bicarbonates with the cations present.

Proteins include egg albumen, milk solids and gelatine from animal tissue. They usually have large molecular structures consisting of amino acids which contain carbon, hydrogen, oxygen and nitrogen, and sometimes sulphur and phosphorus. The sewage treatment process breaks them down into smaller molecules called polypeptides and eventually to amino acids, fatty acids, various nitrogen compounds such as ammonium salts and nitrates, as well as organic phosphates and sulphides.

Fats (including waxes and oils) are built up from glycerol and fatty acids. An example is palmitin which is also known as palm oil because of its presence in the pulp of the fruit of palm trees:

$$
\begin{array}{lll}
\text{CH}_2\text{OH} & & \text{CH}_2\text{OCOC}_{15}\text{H}_{31} \\
| & & | \\
\text{CHOH} \quad + 3\ \text{C}_{15}\text{H}_{31}\text{COOH} \rightarrow & \text{CHOCOC}_{15}\text{H}_{31} \quad + 3\text{H}_2\text{O} \\
| & & | \\
\text{CH}_2\text{OH} & & \text{CH}_2\text{OCOC}_{15}\text{H}_{31}
\end{array}
$$

glycerol   palmitic acid   palmitin

The bacteria in the sewage reverse the reaction and the fatty acids formed, such as palmitic acid, are then further degraded into simpler fatty acids, carbon dioxide and water.

Soaps are manufactured from fats by reacting them with sodium or potassium hydroxide to yield glycerol and the sodium or potassium salts of fatty acids (this process is called saponification). Examples of soaps are sodium palmitate $C_{15}H_{31}COONa$, and sodium stearate $C_{17}H_{35}COONa$.

Detergents are synthetic organic compounds made from by-products of the oil industry. They were produced because of the demand for a better product than soap for washing clothes. Before their invention, scum formation in hard water areas gave rise to problems because of the reaction of the soap with calcium and magnesium in the water to form insoluble stearates and palmitates. In effect, the soap was acting as a water-softening agent. The soap first reacted with the calcium and magnesium in the water and only after the insoluble salts were formed could a lather be produced and the clothes cleaned. The problem was that the insoluble calcium soap scum got trapped in the fibres and so clothes often didn't look properly clean compared with today's 'shining brightness'!

With the introduction of synthetic detergents, the cleaning agent didn't soften the water because they did not form insoluble salts with the calcium and magnesium. Other substances were added called 'builders' which soften the water by a process of extracting and binding the calcium and magnesium ions thus removing them from the wash water solution and reducing the hardness of the water.

The most common builder is sodium tripolyphosphate (STPP) and it is a major component in detergent powder. It has the advantage of not only forming strong and soluble complexes with calcium and magnesium but also disperses the dirt in washing solutions. It also makes the wash water slightly alkaline which improves the efficiency of the detergent.

Other materials added to detergent formulations are enzymes, whiteners, dyes, optical brighteners, perfumes and carboxymethyl cellulose. You will see some of these chemicals listed on the detergent containers. As an example, the contents of a typical 'concentrated' washing powder are:

Greater than 30 per cent	sodium tripolyphosphate
15–30 per cent	sodium carbonate
5–15 per cent	sodium perborate
	non-ionic detergents
	sodium disilicate
Less than 5 per cent	amphoteric detergent
	tetraacetylethylenediamine
	soap
	zeolite
	carboxymethylcellulose
	phosphonate
	perfume
	foam suppressant
	optical brightening agents

All of these chemicals enter the sewerage system and the treatment works each time the washing machine is used for laundering clothes.

One of the problems caused by synthetic detergents is that the STPP breaks down in the sewage treatment process to simple phosphates which are then discharged into the river. Phosphate is an essential nutrient for the growth of algae and aquatic plants, and too much causes eutrophication (nutrient richness) which results in excess algal growth (see Chapter 3).

Inorganic substances
Inorganic substances in sewage originate from both domestic and industrial wastes.

In domestic sewage, the inorganic components are mostly formed from the organic waste. Thus the nitrate and ammonium salts come from the breakdown of proteins which are present in waste food, faeces and urine. Sulphates and phosphates also come from this source. Sulphur is one of the elements in proteins but it is also present in some foods and drinks, for example sulphite is added to wines as an antioxidant. Detergents are a major source of both these anions because one of the commonest detergents is sodium dodecyl benzenesulphonate, which ultimately breaks down to carbon dioxide and sulphate ion. We have already seen that much of the phosphate in the sewage effluent is derived from the builders used in detergent formulations. It has been calculated that, in untreated sewage, 40 per cent of the phosphorus originates from detergents whilst 44 per cent is in our excreta; the remainder comes from household cleaners and from industrial sources.[3]

Bicarbonates and carbonates are formed from the carbon dioxide in the sewage after the breakdown of carbohydrates and other organic compounds. Chloride is present in human urine at a concentration of about 1 per cent. This ion is not affected by the micro-organisms so its concentration is altered mainly by dilution.

Industrial wastewater can contain a variety of inorganic compounds depending on the type of industry. Phosphates are added to water to prevent scaling in boilers whilst chlorine is present in the waste from laundries, paper mills and textile bleaching. Metals at various concentrations originate from metal works such as chromium from plating, silver from photographic developers, lead, cadmium and mercury from battery manufacturers, copper from wire making, iron from castings and can makers and so on. Industries also use a range of alkalis and acids, so the wastewater can contain sulphates, nitrates, phosphates and chlorides as well as sodium, potassium and calcium.

Many of the inorganic constituents in industrial sewage are poisonous to the bacteria that purify the sewage before discharge into the river or sea. Many metals are toxic, such as zinc, copper, chromium and cadmium, whilst excessive amounts of acids or alkalis can diminish the purifying efficiency of the micro-organisms. The amount of toxic waste entering sewers has to be controlled by the sewage authorities. They have 'trade effluent' inspectors on their staff who visit factories and industrial premises to check on the wastes that are put into sewers. Sometimes a waste can be too hazardous for discharge into the sewer and the trade effluent inspector can insist on it being pre-treated at the factory to a standard which will not be harmful to the sewage treatment process.

Quality of effluents from sewage treatment works and from industries

Before any wastewater can be discharged into the river or the sea, permission to do so must be obtained from the organizations that control pollution. As mentioned in Chapter 1, in England and Wales this is the Environment Agency (EA), in Scotland, the Scottish Environment Protection Agency (SEPA) and in Northern Ireland, the Environment and Heritage Department of the Department of the Environment (the addresses of these Agencies are given at the back of this book). These organizations were set up by government to prevent pollution of the environment – not just water but also the air and the land. They also control radioactive wastes.

The Agencies control the amount of waste that is released by a licensing system. The discharger makes an application for 'consent to discharge' and the Agency usually gives its permission based on what are known as 'consent conditions'. Answers must be given to a number of questions. If the effluent is going into a river, the following information is required:

What is in the wastewater?
Where is it being discharged?
What is the flow rate of the river, particularly in summer when the river flow is low and the effluent is not diluted so much as in winter?
What is the flow rate of the effluent?
What is the river going to be used for downstream, for example, is it going to be extracted for water supply?

If the discharge is going into the sea, the Agency will ask:

Where is the discharge entering the sea?
What are the currents in the area for the different states of the tide?
Are there any beaches nearby that people use for bathing in the summer?
Is the sea used for shellfish growing or for fish farming?

Once the answers to these questions have been obtained, the Agency has to decide what consent conditions to apply to the effluent. It has to achieve a balance between, on the one hand, using the rivers and the sea to get rid of our waste (it has to go somewhere!) and, on the other hand, making sure that the effluent doesn't pollute the water. The assumption is made that the rivers and the sea can cope with a certain amount of waste, but it is a matter of getting the balance right.

As an example, let's assume a new electronics factory wants to discharge its effluent into a river that is popular with anglers because it has many trout in it. The factory makes an application to the appropriate environment agency for a consent to discharge and says that its waste-water contains copper. The Agency has to decide on the maximum amount of copper it will allow to be present in the discharge so that it will not harm the trout. It needs information on the effects that copper has on trout and so it looks at all the scientific papers and reports that are available on this subject. Small amounts of copper are essential to life, but large doses are toxic, especially to plants. A complicating factor is that the toxic effect varies according to the hardness of the water. In hard-water areas such as south-east England, the calcium in the water reduces the toxic effect of the copper. So if the effluent is discharging into a stream in this area the Agency can allow more copper to be discharged than if the factory is being set up in parts of Wales or Scotland where the water is much softer and lacking in calcium.

One of the pieces of information that is available for deciding the consent conditions is the toxicity of copper to fish. This is usually expressed as the 96-hour LC_{50}, which is the concentration of the substance that will kill half the fish (the 'lethal concentration') in four days.

For copper the 96-hr LC_{50} to rainbow trout is:

Concentration (µg/l)	Water hardness (mg $CaCO_3$/l)
20–100	14–45
500–1000	200–300

From these figures you can see that copper is much more poisonous in soft water than in hard water. This effect is summarized in Figure 3.

This information tells you only the toxic level of copper, but the Agency has to allow copper to enter the river and make sure the fish are not affected. Another piece of information is that if trout are in water (hardness of 45 mg/l) with a concentration of copper of 10 µg/l, they show increased coughing, so even this concentration affects them adversely.

From these and many other pieces of information, 'safe' concentra-

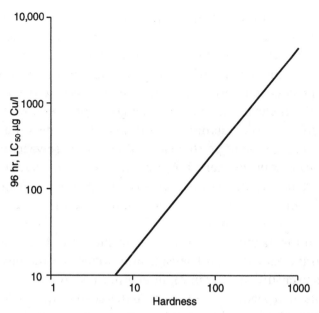

Figure 3. Dependence of copper toxicity on water hardness
Source: G. Mance, Pollution Threat of Heavy Metals in Aquatic Environments,
Elsevier, London, 1987

Table 2. Safe concentrations for copper (µg/l) in different
hardnesses of water

Water hardness	Copper concentration
0–50	1.0
50–100	6.0
100–250	10.0
over 250	28.0

tions can be set and those for the UK can be seen in Table 2. Now that
the safe level is known, the Agency can decide what is the upper limit
that can be allowed to be present in the discharge from the electronics
factory. For example, if the wastewater is entering a stream with a water
hardness of 75 mg/l and the effluent is diluted 20 times by the river water,
then the maximum concentration allowed will be 120 µg/l (0.12 mg/l),
but if the effluent is going to a river with a water hardness of 300 mg/l
then the limit will be 560 µg/l (0.56 mg/l).

This was an example of how the regulatory authorities decide how
much waste a river can absorb. The great majority of industries, though,

dispose of their wastewater through the sewerage system and it is then treated at the sewage treatment works. However, as we learned in the section on the sewage treatment processes, the purification of sewage depends upon the activity of micro-organisms. In other words, it's a biological process and, just as the environment agencies try to preserve the life in the rivers by controlling what goes into them, so the sewage authorities restrict the amount of toxic waste that enters the sewerage system in order to maintain the bacterial life in the sewage works. The sewage works, however, can absorb waste more effectively than rivers, so the standards are not as strict. This is why it is often cheaper for an industry to put its waste into the sewer rather than treat it to the standard required for going into a river.

Although the example above was about the control of copper pollution, the principal concern of sewage authorities and the environment agencies is to ensure that organic matter is purified to a level that will not affect the life in receiving waters. Later in this chapter we will learn about the effect that biodegradable organic matter has on water chemistry. The amount of organic matter in sewage and sewage effluent is measured by a special test called the biochemical oxygen demand or BOD test. The BOD is a measure of the amount of oxygen used up by micro-organisms as they decompose the degradable organic matter that is present. The test involves measuring the dissolved oxygen in the sample initially and then after it has been incubated in the dark for five days at 20°C. You may wonder how these parameters were decided! The test was devised in 1912 by the Royal Commission on Sewage Disposal and is supposed to simulate the discharge of organic waste into a river in summer time; the five days is an estimate of how long the effluent might be in the river as it makes its slow journey to the sea.

The BOD of untreated sewage varies from place to place but is about 300 mg/l, whilst the 'consent condition' for most sewage effluents is that the BOD should be less than 20 mg/l. In other words, sewage treatment has to remove over 90 per cent of the organic matter. The majority of sewage works achieve this. Crude sewage, as it arrives at the works throughout the day and night, varies both in the flow rate and in the BOD content, and these variations reflect our day-to-day activities. The greatest change takes place at mid-morning when the sewage from all our wakening activities (washing, using the toilet, having breakfast) goes into the sewer. There is another minor peak flow rate in the early evening when we've returned home from school and work.

Assuming that the sewage treatment works is operating efficiently and meeting the standard set by the regulatory authority, we can predict the main constituents in the effluent and look at their concentrations both in the effluent and the receiving water.

Effluents from well-operated works

As we saw earlier when we considered the chemistry of sewage treatment, a modern sewage works will oxidize carbohydrates to carbon dioxide, some of which forms carbonates and bicarbonates. The proteins are broken down to amino acids, ammonium salts and sulphates, whilst sulphur and phosphorus compounds are converted to sulphates and phosphates. The ammonium salts that are formed are often further oxidized by bacteria to nitrates. To summarize, the oxidation products of sewage treatment occur as follows:

Organically combined carbon $\rightarrow$ CO_2, HCO_3^- bicarbonates and CO_3^{2-} carbonates

"   nitrogen $\rightarrow$ NH_3 ammonia and NO_3^- nitrates

"   sulphur $\rightarrow$ SO_4^{2-} sulphates

"   phosphorus $\rightarrow$ PO_4^{3-} phosphates

The effluent will be a clean-looking water with a small amount of suspended matter in it, the BOD will be less than 20 mg/l and it will have the above listed salts in solution as well as traces of detergent.

The results of chemical analysis of a typical good quality effluent and its effect on the chemistry of the receiving water are shown in Table 3. This set of analytical results shows that there is a small amount of organic matter in the effluent and this slightly increases the BOD of the river water. As this organic matter is broken down by the naturally occurring aerobic bacteria in the water, the dissolved oxygen is slightly reduced. If a sample was taken further downstream, the dissolved oxygen level would recover because more oxygen would dissolve in the water from the atmosphere. All the other parameters also show increases of various sizes when comparing the chemical quality of the river upstream and downstream of the discharge point. However, there is nothing in the set of results of the downstream sample to suggest that the quality is adversely affected and it is likely that the aquatic life from both river sites will be similar. Some of the tests listed can be readily carried out in the school laboratory

Table 3. Chemical analysis of a good quality sewage effluent and its receiving waters

Chemical determinand	Concentration (in mg/l, except pH and conductivity)		
	River water upstream	Effluent	River water downstream
BOD	3.3	12.0	3.6
Suspended solids	12.0	20.0	16.0
Dissolved oxygen	10.3	7.1	10.0
Permanganate value*	4.4	6.0	5.1
Temperature	12.0	15.0	12.0
Ammonium nitrogen	0.6	13.5	0.9
Nitrate nitrogen	1.3	0.7	1.8
Detergent	0.0	0.8	0.1
Phosphate phosphorus	0.2	1.0	0.4
Alkalinity (as $CaCO_3$)	60.0	150	70.0
Chloride	14.0	62.0	23.0
pH	7.1	7.5	7.2
Conductivity	393	701	434

* The permanganate value is another measure of organic matter but it is a chemical oxidation rather than a biochemical one. The details of the test are given in Chapter 12.

and will be described later in the book. Thus it is possible to carry out a project and assess the impact that an effluent is having on a receiving stream.

Effluents from poorly operated works

Although there has been a great improvement in the treatment of sewage in the last 10–15 years because of investment in new plant and equipment, there are still some sewage works where the wastewater is not adequately treated. This may be because they are old and the equipment is inefficient, or they may be overloaded, i.e. they are receiving more sewage than they were designed to purify. In these situations, the effluent that enters the receiving water still contains many of the substances that were present in the incoming sewage. Some have been only partially broken down by the bacteria so the effluent contains high concentrations of organic matter.

We saw in the description of the operation of sewage works how the presence of oxygen is important for the treatment of sewage. It enables the sewage to be degraded and oxidized by aerobic bacteria (which require free oxygen for living). If the sewage works is overloaded, broken down or operating inefficiently, then there may not be sufficient oxygen for the aerobic bacteria and the organic matter is then broken down by anaerobic bacteria (those that can live in the absence of oxygen). Anaerobic bacteria obtain their oxygen from the breakdown of salts such as phosphates and sulphates. The products of this degradation are often foul smelling and occur as follows:

Organically combined carbon → methane
      "   nitrogen → amines and ammonia
      "   sulphur → sulphides and organic sulphur
                                compounds
      "   phosphorus → phosphines and organic
                                phosphorus compounds

The effects of a poor quality effluent on a receiving stream can be very marked and it may take many kilometres for the river to recover from the pollution. This is because the river downstream acts like an extension to the sewage works in that the naturally occurring bacteria and protozoa oxidize the organic matter present in the sewage effluent and, as they do so, they utilize the dissolved oxygen in the river water. However, the amount of oxygen which is dissolved in water is limited, for example at 10°C, water contains only 11.3 mg/l of dissolved oxygen whereas air contains nearly 300 mg/l. The dissolved oxygen may therefore be quickly used up and it takes time for fresh oxygen to dissolve in the water from the air, especially if the water is slow flowing. If the dissolved oxygen is depleted, then the substances formed in anaerobic conditions are present in the water and many of them are poisonous to the river organisms. However, as the river flows over rocks and waterfalls, the oxygen levels are restored and the organic matter is degraded to the extent that aquatic life is returned to normal. This process is called self-purification.

The chemical results for a poor quality effluent and for samples taken upstream and downstream, are shown in Table 4. A poor quality effluent usually contains high concentrations of ammonia as well as organic matter. As it flows downstream, this ammonia is oxidized by bacteria to nitrate. This process is called nitrification and also utilizes dissolved oxygen for the oxidation process. The overall effect on the river chemistry downstream of a poor quality discharge is illustrated in Figure 4.

Table 4. Chemical analysis of a poor quality sewage effluent and the receiving waters

Chemical determinand	Concentration (in mg/l, except for pH and conductivity)		
	River water upstream	Effluent	River water downstream
Suspended solids	11.0	120.0	46.0
BOD	2.8	105	5.4
Permanganate value	9.4	31.6	11.0
Dissolved oxygen	11.7	0.0	6.4
Ammonia nitrogen	0.3	32.0	7.8
Nitrate nitrogen	0.9	1.7	0.5
Detergents	0.1	4.5	0.9
Phosphate phosphorus	0.3	3.7	1.2
Alkalinity	65.0	170.0	115.0
Chloride	4.0	75.0	22.0
pH	7.0	7.7	7.7
Conductivity	143	820	370

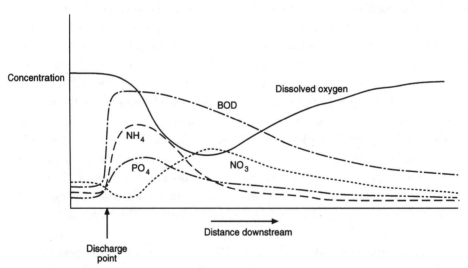

Figure 4. Changes in the chemistry of river water downstream of a poorly treated sewage effluent

Source: The Biology of Pollution, *Macan, Hodder. Reproduced with permission*

Sewage treatment in East Kilbride, Scotland: a case study

East Kilbride is located about 15 km south-east of Glasgow in an area of high ground. It is one of Scotland's 'new towns' and was built to take the overspill of Glasgow's population. Originally it was a small village of a few hundred people, but it expanded very rapidly from 1944. Now it is one of Scotland's largest towns with a population of about 85,000 people. As it was built on a hill, the drainage from the town flows in two directions from the high point. In the west side of the town, the sewage flows to Philipshill sewage treatment works (STW), whilst the eastern side's wastewater goes to Allers STW.

The effluent from Allers STW flows into a small tributary of the River Clyde called the Rotten Calder Water (so-named because of the two main streams in the catchment, the Rotten Burn and the Calder Water, and not because of its condition!). When the town was small, the effluent made little impact on the quality of the river, but as it expanded, the flow rate of the effluent made an increasingly large contribution to the river flow. In 1996, in the long dry spell in July and August, the river flow diminished but the effluent flow remained the same. Downstream of the effluent discharge point, the flow rate of the river was composed of 75 per cent sewage effluent.

Although the Allers STW was modernized and expanded to keep pace with the amount of sewage it had to purify, nevertheless the effluent quality was not good and the Rotten Calder Water started to live up to its name. The results in Table 5 are for some selected determinands of the river upstream and downstream of the effluent point.

The effluent was clearly having a bad effect on the river, and the variety of aquatic life was very reduced downstream of the effluent. The trout

Table 5. Chemical quality of the Rotten Calder Water upstream and downstream of the effluent point of Allers STW, summer 1996

Determinand	Concentration (in mg/l)	
	Upstream	Downstream
BOD	2.5	4.0
Dissolved oxygen	8.8	7.6
Ammonia nitrogen	0.28	8.6
Nitrate nitrogen	1.2	1.7
Chloride	60.0	77.0

that used to live in the river had long since disappeared and there were complaints about the smell of the river in summer months.

The problem for the local authority which manages the STW was that, however good the treatment given to the sewage before discharge, there would always be a lack of clean river water to dilute the effluent. It was eventually decided that the only solution was to divert the effluent away from the Rotten Calder Water and discharge it where it would receive more dilution. This work started in 1995 and was completed at the end of 1996. Now the effluent flows into the River Clyde where it is diluted at least 50 times, even when there are low summer flows. The effects of this on the quality of the Rotten Calder have been dramatic as shown in the results for the survey carried out at the beginning of 1997 in Table 6.

Table 6. Chemical quality of the Rotten Calder Water upstream and downstream of the former discharge point of Allers STW, February 1997

| Determinand | Concentration (in mg/l) | |
	Upstream	Downstream
BOD	1.6	2.3
Dissolved oxygen	11.8	11.5
Ammonia nitrogen	0.08	0.05
Nitrate nitrogen	2.4	2.4
Chloride	29	32

Now that the major source of pollution has been removed from the river, it is expected that the trout will return and the variety of aquatic life will increase.

Notes

1. *Activated Sludge Treatment*, Chartered Institution of Water and Environmental Management, London, 1997.
2. 'A Study of WC derived Sewer Solids', Friedler E., Brown D.M. and Butler D., *Water Science and Technology* 33, 9, 1996, 17–24.
3. *United Kingdom Soaps and Detergent Industries Association, Detergent Phosphate and Water Quality in the UK*, SDIA, Hayes, Middlesex, 1989.

Further reading

B.J. Alloway and D.C. Ayres, *Chemical Principles of Environmental Pollution*, Blackie Academic, London, second edition, 1997.

R.M. Harrison, *Understanding our Environment: An Introduction to Environmental Chemistry and Pollution*, Royal Society of Chemistry, London, second edition, 1992.

J.R. Dojlido and G.A. Best, *Chemistry of Water and Water Pollution*, Ellis Horwood, Hemel Hempstead, 1993.

3. Eutrophication

The word 'eutrophication' comes from the Greek *eutrophos* which means well nourished. It is applied to water that is enriched with nutrients, mainly phosphorus and nitrogen compounds, which encourage the growth of abnormally large number of algae and aquatic plants. The extent of nutrient enrichment of water (its 'trophic' state) is described by different prefixes, graded from ultra-oligotrophic water, which is very deficient in nutrients, through oligotrophic, mesotrophic and eutrophic, to hypertrophic, which has a great excess of nutrients.

The problem of eutrophication mainly applies to still water such as that in lakes, ponds and canals. This is because the static water allows sufficient time for the algae to grow and multiply, whereas in rivers they are constantly being moved and swept downstream. The excess nutrients in rivers, however, do encourage the growth of water weeds and attached algae, and the river can become choked with the mass of plants.

In modern times, the amount of eutrophication has increased rapidly as a result of human activities. Large quantities of nutrients are discharged into surface waters and the sea from industry and from sewage treatment works. They also originate from agriculture, especially from the application of fertilizers to crops and from the spreading of animal wastes onto fields.

Aquatic algae require four main components for their growth: carbon dioxide and sunlight for photosynthesis, warmth and nutrients. So in eutrophic waters in summer time, there is often a prolific growth of algae on the water surface. This causes problems in the lake or pond because:

- the water becomes turbid with tens of thousands of algal cells in each millilitre
- the sunlight cannot penetrate the deeper water because of the density of the algae
- the wind can blow the algae onto the shore line where they form

unsightly and smelly deposits as they rot
- when the algae die off at the end of the summer, the decomposition process uses up dissolved oxygen from the water and kills off other aquatic life.

The main nutrients needed by the algae and other plants for growth are nitrogen, phosphorus, potassium and silicon. There are also many trace elements required, such as cobalt, iron, manganese, molybdenum, vanadium, etc. In fresh waters, the problem of excess algae (algal blooms) is caused mainly by too much phosphorus in the water, whilst in the sea it is often caused by excess nitrogen.

Freshwater eutrophication

Algal blooms have increased markedly in many freshwater lakes and ponds; they also occur in very-slow moving water, in canals and rivers such as the Norfolk Broads. The problem can be attributed to the increasing amount of phosphorus entering the affected waters. We have already seen that phosphates are a major component in modern detergents, but they are also present in human sewage, animal excreta, industrial effluents and agricultural fertilizers.

In most freshwater lakes and rivers, phosphorus is known as a limiting nutrient. In other words, all the other nutrients required by the plants and algae are in excessive amounts and the growth rate is decided by the amount of phosphorus present. If phosphorus is added to oligotrophic water, algal and plant growth is triggered and the trophic status shifts towards mesotrophic. Figure 5 shows the relationship between the concentration of phosphorus in the water and the growth of algae.[1] The algal growth is measured by the concentration of chlorophyll which is the green pigment in the algal cell.

When phosphorus enters a lake from either a discharge or in an inflowing stream, it can be present in a number of different forms:

- soluble inorganic phosphate
- soluble organic phosphorus compounds
- phosphorus absorbed onto suspended particles in the water (particulate phosphorus)

When present as particulate phosphorus, these particles slowly settle to the bottom of the lake and become part of the sediment. Sometimes

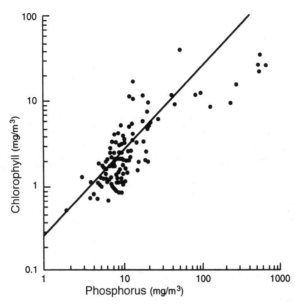

Figure 5. The relationship between the amount of phosphorus and the
development of algae, as measured by the concentration of chlorophyll
Source: Phytoplankton Ecology, *G.P. Harris, 1986. With kind permission
from Kluwer Publishers*

the particles are eaten by fish or microscopic animals. The phosphorus
then passes through their bodies and is excreted in their faeces. This movement of the element in different sectors of the water is known as
phosphorus cycling and is summarized in Figure 6.

The algae require soluble inorganic phosphate for their growth and, as
we see from Figure 6, this can be present in the inflowing stream but is
also formed from the phosphorus cycling within the lake. For this reason,
the nutrient status of water quality is classified by the concentration of
total phosphorus in the water.[1]

Table 7. Classification of standing waters	
Trophic category	Concentration of total P (µg/l)
Ultra-oligotrophic	<4
Oligotrophic	4–10
Mesotrophic	10–35
Eutrophic	35–100
Hypertrophic	>100

Source: Eutrophication of Waters: Monitoring, Assessment and Control, *Organisation for
Economic Cooperation and Development, OECD, Paris, 1982*

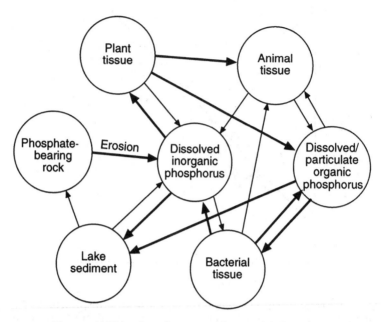

Figure 6. The phosphorus cycle in a freshwater lake
Source: Ecological Effects of Wastewater. *Second edition. E.B. Welsh. With kind permission from Kluwer Publishers*

Oligotrophic lakes and ponds are usually found in upland areas where the surrounding soils are lacking in minerals (usually because they have not been treated with fertilizers by farmers). Eutrophic lakes are found in lowland areas; the surrounding fields may be well fertilized to encourage crop or grass growth and the inflowing stream may contain effluent from the sewage works of a nearby community.

In the summer months, the rapid growth of algae in a eutrophic lake removes the soluble phosphorus and nitrogen and the concentration of these nutrients in the water declines. They are replenished in the winter by inflowing water or by resuspension from the sediment. The annual cycle of algal growth and changes in phosphorus concentration is illustrated in Figure 7.

There is an additional factor to consider in the eutrophication of lakes and the sea in the temperate climates of the southern and northern hemispheres, and that is stratification.

In winter months, the air and water are cold, and the air is often colder than the water. Also, it tends to be windier in winter months than in the summer. The result of these weather factors is that the temperature of the water in winter is almost the same at whatever depth you sample it.

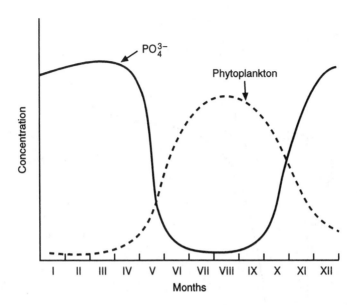

Figure 7. Changes in phosphorus and phytoplankton in a lake over a year
Source: The Chemistry of Water and Water Pollution. *Dojlido and Best, 1993. E & F.N.*
Spon. Reproduced with permission

It also has the same density because it is well mixed throughout by the wind.

In late spring and summer, the water is heated by the sun and, in calm conditions, the surface layers get warmer and less dense than the deeper water. If the calm weather persists, the water becomes stratified. This means that the surface layers do not mix with the deeper layers because there is a marked difference in temperature and, consequently, density between the two parts. This is illustrated in Figure 8.

A case history of a eutrophic lake

Strathclyde Park Loch is situated between Hamilton and Motherwell in South Lanarkshire, Scotland. It is an artificial loch and was created from an area of derelict and subsided ground which was a coal mine at one time. The original pond in the subsided ground was extended and contoured to make an attractive amenity lake and became the central feature of Strathclyde Country Park.

The loch is very popular in the summer months with anglers, dinghy sailors, canoeists and wind surfers. Unfortunately it also suffers from algal

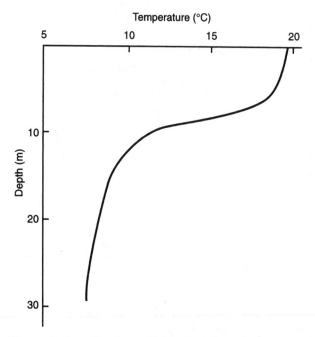

Figure 8. Stratification of lakes in calm summer weather

blooms in the summer because the water is eutrophic. The loch is filled by the South Calder Water and the overflow goes to the River Clyde.

The South Calder Water receives the effluents from three sewage treatment works at different places along its length, and all of which contribute nitrates and phosphates to the river and these eventually enter Strathclyde Park Loch. Table 8 illustrates the concentration of these nutrients in different seasons.

From these results it can be seen that, in winter months, the concentrations of nitrate and phosphate for the sampling points at the inlet and outlet to the loch are almost the same because the lack of sunlight and the cold water are unsuitable for algae to develop. By contrast, in the summer months, there is a marked reduction in the concentrations of the nutrients at the outlet in comparison to the inlet because they are being used up by the fast-growing algae within the lake. The overall concentrations of the nutrients in the samples taken in the summer months are higher than the winter values because the larger flows in the river in the winter dilute the effluents to a greater degree.

The excess algae in the loch in summer make the loch look very green and turbid – on windy days, the wave action blows the algae into a froth

Table 8. Concentrations of nutrients in the inlet and outlet of Strathclyde Park Loch in winter and summer (mg/l)

		Inlet	Outlet
1995			
Winter	Phosphate phosphorus	0.19	0.12
	Nitrate nitrogen	1.68	1.70
Summer	Phosphate phosphorus	0.42	0.16
	Nitrate nitrogen	1.89	0.53
1996			
Winter	Phosphate phosphorus	0.22	0.22
	Nitrate nitrogen	1.74	1.63
Summer	Phosphate phosphorus	0.70	0.41
	Nitrate nitrogen	2.35	1.22

which accumulates on the shore line where it is a hazard to people and their pets. An example of algal froth is shown in Cover Illustration 2.

One particular type of algae, known as blue-green algae, is a danger to the public and to people who use the water. This is because the algae produce a poison as they photosynthesize, and this enters the water. People or animals that swallow the water can suffer from a variety of illnesses ranging from ear infections and sore throats to diarrhoea.

The occurrence of toxic algae is regularly monitored by staff from the Scottish Environment Protection Agency. The biologist who specializes in identifying algae, an algologist, collects samples of the loch water at regular intervals from spring to autumn and, on return to the laboratory, looks for the presence of blue-green algae. If the numbers exceed a safe level, a warning notice is put up at the lochside warning of the danger and advising people to prevent their dogs from drinking the water.

Figure 9 shows when the blue-green algae appeared in 1993–96 and how long the numbers exceeded the safe level. During one of these periods, the rowing events for the Commonwealth Games were due to be held but they were postponed because of the presence of the toxic algae.

Strathclyde Park Loch is not the only freshwater lake to suffer from poisonous algae. In 1989, sheep and dogs died after drinking algae-infected water at Rutland Water whilst soldiers who had been swimming in Rudyard Lake in Staffordshire later became ill.

In the example described above, of Strathclyde Park Loch, the nutrients that gave rise to the algal problem originated from the sewage works

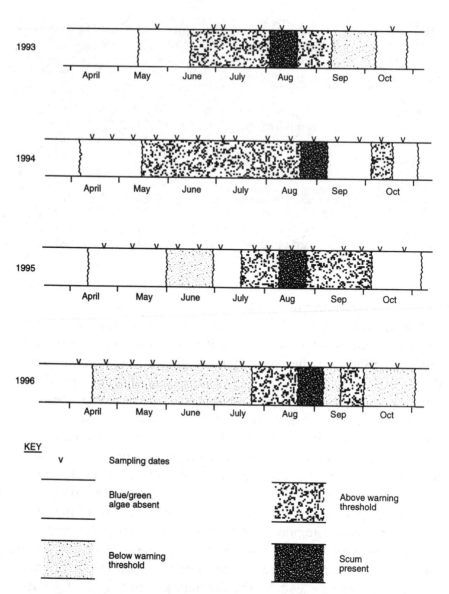

Figure 9. The appearance and intensity of blue-green algae in
Strathclyde Park Loch, 1993–96

that discharged into the feeder river. In other lakes, the nutrients can
enter the water in drainage from agricultural land. Farmers add fertilizers
to increase the growth rate or size of their crops (see Chapter 4). This
particularly applies to arable (vegetables) farming, where the amounts
added to the crops are greater than for stock (animal) farming. Fertilizer

usually contains three nutrients, nitrogen, phosphorus and potassium, usually called 'NPK'.

Not all the nutrients stay in the soil and some are washed into streams by rain. The amount of the nutrients that enter the streams depends on many factors such as the type of soil and crop, the slope of the land, how much fertilizer was applied, etc. In general, the nitrate is more readily washed out (leached) from the soil than phosphate whilst the potassium is most strongly attached to the soil. The loss of nitrate can vary from 5 to 50 per cent of the amount applied whereas the phosphate losses are in the range 0.1 to 5 per cent.

However, although the amount of phosphate lost is relatively small, after a rain storm it can still reach a concentration in a stream of 3 mg/l which is well in excess of the 0.1 mg/l (100 µg/l) which gives rise to hyper-trophic water. It has been estimated that leaching of phosphate from agricultural land can account for 55 per cent of the total amount of phosphorus that enters fresh waters.

Fertilizers are costly and farmers should not waste them by applying too much or at the wrong time. In an effort to reduce the amount of phosphorus entering rivers and lakes, farmers are now given advice by the Agricultural Advisory Service on how much fertilizer their fields require for the appropriate crop and when is the best time to apply it. They are also encouraged to leave a so-called buffer strip next to any rivers so that fertilizers are not spread onto the ground immediately adjacent to the water. Figure 10 shows an ideal buffer strip which protects a stream from receiving excess nutrients from the cultivated fields.

There has been an interesting new development in the control of algae in freshwater lakes and that is the use of barley straw. It had been noticed, when barley straw had accidentally fallen into a small lake that was usually green with algae in the summer, that excessive growths of algae did not occur during the following summer. Since this accidental discovery, many authorities have been using barley straw to control algal blooms. It is not known yet how this works but it appears that, as the straw decomposes in the water, some substance is released that is toxic to the algae. This algicidal chemical has yet to be identified but it seems to be released after microbial decomposition of the straw; research is still being carried out to find the active ingredient. However, at this stage the findings of this research are not important because the straw is readily available and it is not necessary to obtain anyone's approval to put it into a lake or pond. If the active ingredient is identified, it will take years before

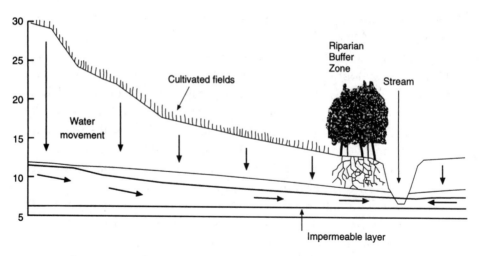

Figure 10. Buffer strip between cultivated fields and a stream

it can be marketed as a treatment chemical because it will have to have the approval of many different organizations involved in checking its toxicity and safety before it is licensed for use. Meanwhile, throughout the UK and elsewhere, problems of excessive algae, particularly blue-green algae, are being controlled by putting either bales or long tubes of netting filled with straw into fresh water early in the spring so as to start the decomposition process by the summer. It has been used with some success, for example, in Strathclyde Park Loch where it seems to have reduced the intensity of the algal bloom.

A more systematic experiment was conducted in the Press Top Reservoir in Derbyshire which had been used for water supply storage but is now a recreational fishery.[2] There was a noticeable decline in the fish stocks in 1990–93 which was blamed on the development of blooms of blue-green algae. In April 1994 the reservoir was treated with barley straw at a rate of 50 g/m^3 and the extent of the algal growth was monitored and compared with earlier years when no straw had been added. The results are shown in Figure 11 and seem to suggest that the straw has successfully reduced the extent of the algal problem.

Eutrophication in marine waters

Much less is known about eutrophication in sea water because it has not been studied as extensively as fresh water.[3] Algal blooms do occur though and can be harmful at times.

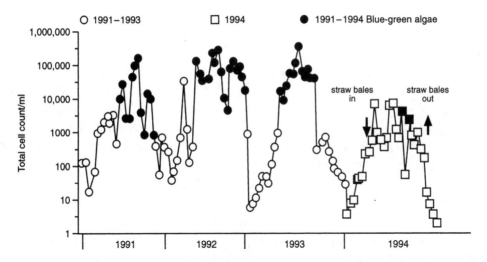

Figure 11. General and blue-green algal numbers in Press Top Reservoir
1990–94

Reprinted from Water Research *30, 2, Everall and Lees, 'Use of barley straw to control general and blue-green algal growth in a Derbyshire reservoir'. With permission from Elsevier Science*

Most coastal towns discharge their sewage into the sea with minimal treatment. Usually, the sewage is collected together at a central discharge point and then piped into the sea some distance from the coast after it has been 'screened' – this means that it has been filtered through a wire-mesh screen that retains all particles above a certain size, usually 6 mm. As a result, the amount of nutrients in the sewage discharged from the coastal communities is greater than that from inland ones because none has been removed by the treatment process.

In the sea, the marine algae have the same requirements as those of fresh water, namely, adequate sunlight and the nutrients nitrate and phosphate. However, one particular type of marine algae (or phyto-plankton, also known as diatoms) also requires silica, SiO_2, in order to make a small skeletal structure (diatoms also occur in fresh water: see Chapter 8 dealing with acid rain).

Phosphate is not in short supply in sea water and, in most situations, the nitrate concentration is the limiting factor for algal growth. Nitrate enters the sea from sewage discharges and from rivers. We learned earlier that nitrate is readily leached from the soil by rain and as much as 50 per cent of the fertilizer applied to the land can end up in the river.

The development of an algal bloom in the sea depends on many factors, but one of the most important is calm weather because this encourages the development of stratification. The still waters allow the algae to multiply but also to become concentrated in a small area.

Just as we learned about the hazards of blue-green algae in fresh water, so there are toxic algae in sea water. One of the commonest is called a 'red tide' because the vast number of brown single-cell organisms (*Gyrodinium aureoleum*) give a distinct red colour to the water. Red tides have been responsible for the deaths of thousands of salmon in fish farms in sea lochs in Scotland, Norway and elsewhere. In August 1996, such a bloom was responsible for the deaths of thousands of marine invertebrates in the west and north of Scotland. Reports were received from Islay, Coll, Tiree and Orkney that dead lugworms, sea urchins and shellfish were washed up on the beaches. It appears that the deaths were caused by decaying algae being washed into shallow waters in these areas and smothering the animals that lived in the sediments.

Another problem caused by toxic marine algae is contamination of shellfish which gives rise to a condition known as 'paralytic shellfish poisoning'. Shellfish such as mussels and scallops are filter feeders. They suck in the sea water surrounding them, extract any edible particles and eject the rest. If there are algae in the water they become incorporated into the shellfish flesh. In many places, shellfish are collected for food but this has to be halted if the shellfish have ingested toxic algae. Eating the infected shellfish can cause paralysis or even death, and such poisoning has occurred every year since 1990 off the North and East coasts of Scotland and England and toxic algal blooms have been reported off the South coast of Wales and in the English Channel.

Notes

1. *The Role of Phosphorus in the Growth of Cladophera*, C.E.R. Pitcairn and H.A. Hawkes, Water Research 7 (1973), 159–171.
2. The use of barley straw to control general and blue-green algal growth in a Derbyshire reservoir, N.C. Everall and D.R. Lees, *Water Research* 30, 2, (1996) 296-276.
3. Marine Eutrophication, *Ambio* (Special Edition) X1X, 3, May 1990.

Further reading

E.B. Welch, *Ecological Effects of Wastewater, Applied Limnology and Pollutant Effects*, Chapman and Hall, London, second edition, 1992.
R.B. Clark, *Marine Pollution*, Oxford University Press, Oxford, third edition, 1992.

4. Pollution from farming

Farming has been revolutionized in the past hundred years. In the 1890s, wheat was sown by hand, weeded by the hoe and harvested with the sickle and scythe, and about a third of the population were employed in some activity related to the growing of food. In 1945, 1 million people worked in farming in the UK, but by 1960, regular farm jobs had fallen to half a million; in 1994, the number had declined further to only 120,000 people. The largest single factor contributing to this change has been mechanization, in particular the replacement of the horse by the tractor. Harry Ferguson's prototype tractor arrived in 1933; by the mid-1950s, tractors outnumbered horses in the UK and now there are about half a million of them. The power of the tractor has increased greatly from Ferguson's prototype: it is estimated that modern farming requires the equivalent of 25 million horses.[1]

One of the consequences of this mechanization is the change in the size of the fields. If a farmer is working in a 2-hectare field with a 3-metre wide implement behind the tractor, one-third of the time is spent culti-vating the ground whilst two-thirds is spent turning the machine around, going round the edges or going to another field. If the field is 40 hectares, however, two-thirds of the time is spent cultivating. For this reason, fields have been enlarged to accommodate the bigger machines and to increase productivity. The bigger fields are created by removing the hedges and fences. In the period between 1978 and 1984, 28,000 km of hedges were removed by farmers and only 3,500 km of new ones planted.

Surprisingly, although farmers have been creating bigger fields to maxi-mize yields, the amount of land used for agriculture in the UK has declined in recent years. In 1980, 19 million hectares were used but ten years later this had been reduced to 18.5 million hectares. One of the reasons for this has been the introduction of the Set Aside Scheme introduced by the European Union. This scheme controls the amount of land used by farmers for growing cereals to prevent over-production. Farmers are given a 'ration' of land that they can use for cereals and any excess is set aside, i.e. it is undeveloped and left uncultivated. There has also been a shift in the type of cereals grown in the UK, with less barley (2.3 million hectares in 1980 down to 1.5 million hectares in 1990) but more wheat (up from 1.4 million hectares to 2 million hectares in the same period). Non-cereal crops such as oilseed rape also increased from 1 million to 1.4 million hectares during those ten years. The proportion of land put over to various types of agriculture in the UK is shown in Figure 12.

As well as the crops there have been changes in the proportion of livestock numbers and there are now greater numbers of sheep and pigs and fewer cattle. These changes for Scotland are shown in Figure 13.

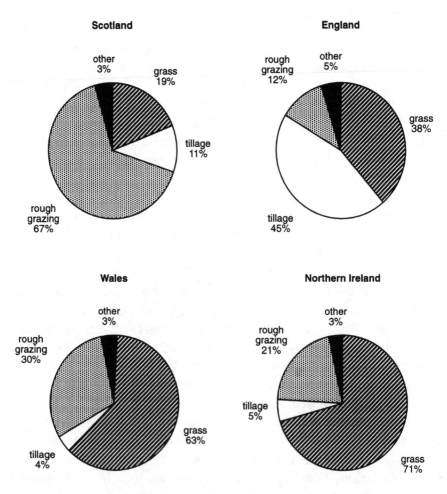

Figure 12. Agricultural areas in different countries in the UK

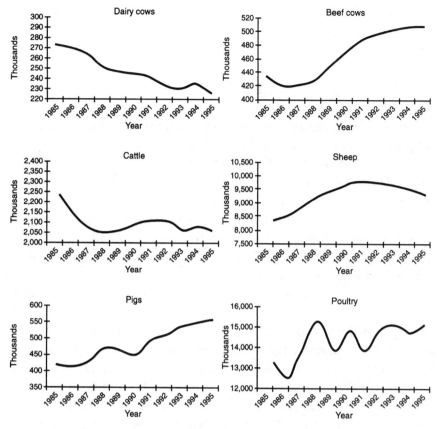

Figure 13. Livestock numbers 1985 and 1995 in Scotland

In the UK as a whole in 1995 there were:

11.7 million cattle
42.8 million sheep
7.5 million pigs
141 million poultry

In order to optimize the yield of the crops sown, farmers apply many chemicals to keep pests at bay and artificial fertilizers to increase the growth rate and the size of the crop. All the livestock produce waste products – this is not a problem with the sheep that are outdoors for most of the time, but cows are now confined to stalls from October until March and their waste products have to be disposed of. In the UK, the annual amount of animal waste which is spread on to land to fertilize the soil amounts to 80 million tonnes.

The fertilizers, pesticides and animal wastes all pose threats to water quality because they are so polluting in different ways.

Fertilizers

Just as we learned in Chapter 3 that algae require nitrogen, phosphorus and potassium for growth, so the same applies for crops, whether it be grass for cattle or wheat for flour. In the early 1990s, 1.5 million tonnes of nitrogen, 400,000 tonnes of phosphate and 500,000 tonnes of potash were applied to fields in the UK. As we also saw in Chapter 3, each of these nutrients has different binding capacity to the soil with nitrate being the least retained. Rainfall can leach nitrates and some phosphates into streams and thus provides perfect conditions for water weeds and algal growths in the receiving waters.

Nitrates are a particular problem in parts of Britain where water is extracted for water supply and it has been contaminated with run-off from cultivated fields. Water, of course, is not the only source of nitrates in our diet. We also get them from our food, with vegetables such as celery, beetroot and lettuce being particularly rich in nitrate. However, if water contains 50 mg/l of nitrate (measured as NO_3, which is equivalent to 11.2 mg/l as N, nitrogen) then this source accounts for 50 per cent of the daily intake of nitrogen. The main concern regarding excess nitrate in drinking water is the development of 'blue baby' syndrome in bottle-fed infants. The nitrate in the feed water is reduced in the body to nitrite and this combines with the haemoglobin in the blood to form methaemoglobin which cannot carry oxygen. If the amount of methaemoglobin in the blood exceeds 10 per cent, the skin takes on a blue tinge because the baby is short of oxygen. This condition is known as methaemoglobinaemia.

Fortunately, this is now a very rare condition in the UK because of the requirement that water supplies must not exceed a maximum concentration of 50 mg/l of nitrate. However, some groundwater supplies (water taken from underground aquifers) exceed this limit and the extracted water has to be blended with water with a low nitrate concentration to reduce it to acceptable levels. The areas most affected are those in which there is most intensive agriculture for arable crops, such as East Anglia, Lincolnshire, Cambridgeshire, Nottinghamshire and Worcestershire. The problem is likely to get worse because it can take many years for the nitrate-contaminated water in the surface soil to percolate through the ground to the aquifer and later be extracted for supplying to homes and businesses. The use of nitrate fertilizer is eight times greater now than 15 years ago, and the contaminated surface water of today may not be extracted for the water supply for another 25 years.[2] The UK government has taken action to reduce the risk. So-called nitrate sensitive areas (NSAs) have been identified where ground water is extracted for water supply and there is agricultural activity in the area. In the NSAs, farming practices have to be altered to reduce the amount of nitrate used on crops.

Pesticides

There are 725 pesticides listed in the *Pesticides Manual*, and of these 450 are approved for use in the UK by the Ministry of Agriculture, Fisheries and Food (MAFF). They are used to control weeds, insects and fungi on crops, animals and fish. In 1993, the amounts of different pesticides used were as follows:

10,700 tonnes of herbicides
6,850 tonnes of fungicides
2,670 tonnes of growth regulators
1,140 tonnes of insecticides
635 tonnes of other pesticides.[3]

Pesticides are usually popularly thought of as being 'bad' chemicals but they protect our food, our homes and the environment (for example, by controlling invasive plants such as bracken and rhododendron). It is estimated that the yield of crops would decline by 40 per cent if they weren't protected by various pesticides. Another statistic is that, world-wide, the loss of crops from pests is estimated at 35 per cent of the total amount which are grown.

The main problem with pesticides is that they can also poison non-target organisms. If pesticides are washed into a stream during rainfall or if the spraying machine accidentally passes over a water course, the pesticides can affect the aquatic life, whether that be the water plants or fish. Fortunately, the problems of environmental pollution from pesticides have decreased markedly in recent years with the introduction of less harmful but effective pesticides and advances in the design of spraying equipment so that less pesticide is used but it still reaches the target pest.

Despite these improvements, there are still concerns about the effects of pesticide spraying on farmland birds. The treatment of cereal crops with pesticides can result in a reduction in the number of invertebrates in and on the soil which are food to songbirds. A survey carried out in the UK showed that in the past 20–30 years the population of certain species of farmland birds had declined significantly, as shown in Table 9. This decline is correlated with the increased use of pesticides: for example, in 1970, 5 per cent of farmland was treated with insecticides, but by 1996, this had

Table 9. Decline in bird numbers on UK farmland in the past 20–30 years

Species	% decline
Tree sparrow	89
Grey partridge	82
Turtle dove	77
Bullfinch	76
Song thrush	73
Lapwing	62
Reed bunting	61
Skylark	58
Linnet	52
Swallow	43
Blackbird	42
Starling	23

Source: The Indirect Effect of Pesticides on Birds, report by the UK Joint Nature Conservation Committee, Natural History Book Service, Totnes, Devon

increased to 90 per cent. The reduction in the bird population has been caused, not by direct poisoning, but by the loss of food for the birds and their young. Some of the birds feed on the insects which are a pest to the crop, whilst for other birds the use of herbicides kills off the weeds and their seeds that they eat. Other factors accounting for the loss are the decline in spring-grown cereals, the decreasing number of mixed farms and the reduction in hedges and ponds in farmland.

In the past, many pesticides were based on organo-chlorine molecules such as DDT (dichlorodiphenyl trichloroethane), Dieldrin and Lindane (see Figure 14).

DDT's insecticidal properties were not discovered until 1939 and thereafter it was used very successfully to control malaria by killing off the mosquitoes which carried the disease. DDT was also used to protect many crops from insect damage. To be effective, pesticides should have the following properties:

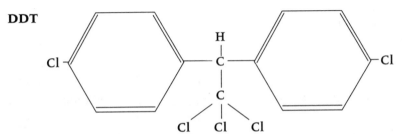

DDT

30,000 tonnes manufactured in 1945
200,000 tonnes manufactured in the late 1950s
1,600,000 tonnes manufactured in 1964
Banned in 1972 in the USA, and in 1984 in the UK

DDD

DDE

Figure 14. Structure of DDT and selected organo-chlorine pesticide molecules

- high toxicity to pests
- low toxicity to other organisms, especially humans and aquatic organisms
- adequate stability so that they complete the task of killing the pest before they degrade, and
- ability to degrade so that after completing their task they disappear into the environment without causing harm.

The organo-chlorine pesticides met most of these criteria but they did not degrade easily. For example, DDT takes between 4 and 30 years to degrade by 95 per cent, and Lindane takes 3–10 years. Another problem was that they bioaccumulated, in other words they were preferentially absorbed into the bodies of birds, animals and fish at much greater concentrations than in the environment because they were more soluble in fats and oils than in water. An example of bioaccumulation is shown in Figure 15.

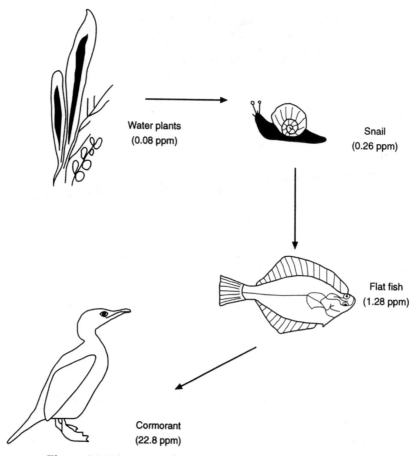

Water plants
(0.08 ppm)

Snail
(0.26 ppm)

Flat fish
(1.28 ppm)

Cormorant
(22.8 ppm)

Figure 15. Bioaccumulation of an organo-chlorine pesticide
in the food chain

The pesticide is washed by rain into the water where it is adsorbed on to aquatic plants. These are grazed by snails which bioaccumulate the DDT because it is preferentially absorbed from the plant into the fatty tissue of the snail. Further bioaccumulation takes place as the snail is eaten by the fish and the fish by the cormorant. If the fish was eaten by a person instead of the bird, then the DDT would be absorbed into the fat of that person. That is why each of us still has traces of DDT and other organo-chlorine compounds in our bodies. In the 1960s, samples of human fatty tissue from different parts of the world were analysed for DDT and the results can be seen in Table 10. The variation depended on how much DDT was used in the country and the food that was eaten.

Table 10. Concentration of DDT in human fat	
Country	Concentration (µg/kg)
England	0.7
Hungary	5.7
Italy	10.6
India	16.0

Source: C.H. Edwards (ed.), Environmental Pollution by Pesticides, Plenum Press, London, 1973

It eventually became clear that these organo-chlorine compounds were causing harm to the environment and the organisms that lived in it. It was noticed that birds were dying or failing to breed properly, that fish and insects were killed when the pesticide drained into streams, and that crop-sprayers and farm-workers were suffering adverse effects after using the substances. The issues were clearly described in *Silent Spring*, the famous book by Rachel Carson, and this had a major impact on pesticide use. DDT and other organo-chlorine compounds were either banned from use or restricted. The USA banned DDT in 1972 and the UK banned it in 1984.

Since then, chemists have devised new and more effective pesticides – in particular, they are much less harmful to the environment because they degrade more rapidly after use. They can also destroy the pest at lower concentrations so less of the substance is used. Some examples of pesticide changes resulting in reduced rates of application are shown in Table 11. Many of the new pesticides are either organo-phosphorus or

Table 11. Changes in pesticide application rates for selected crops, 1983–93				
	1983		1993	
Crop	Pesticide	Application rate (kg/ha)	Pesticide	Application rate (kg/ha)
Oilseed rape	TCA	1.45	fluazifop-P-butyl	0.09
Apples	captan	2.75	myclobutanil	0.05

organo-nitrogen compounds. The absence of the chlorine atoms in their structures makes them more amenable to biological breakdown.

A further new class of pesticides, synthetic pyrethroids, are now used in many applications. They are based on the natural insecticide pyrethrum that can be extracted from the chrysanthemum flower. Scientists have identified the chemical structure of pyrethrum and have made synthetic derivatives of it which are more effective than the natural product. The first synthetic pyrethroid was developed in 1973 and now, over 25 years later, this class of pesticide is one of the most commonly used. Typical uses and crops that are now protected by pyrethroid pesticides are shown in Table 12.

Table 12. Typical uses of pyrethroid pesticides

	Use	Pests controlled
Protection of	Cotton	Boll worms
	Cereals	Aphids
	Carpets and pets	Fleas
	Timber	Wood-boring beetles
Use in	Hospitals	Cockroaches
	Aerosols	Flies and mosquitoes

If you look at the labels on any bottles or aerosols you may have in your house or garden shed for killing garden or house pests you will find that most of them contain synthetic pyrethroids.

One side-effect to the success of the pyrethroid pesticides is that they are very toxic to aquatic life – more toxic in fact that some of the pesticides they have replaced. This can be seen in the use of pesticides to protect sheep from parasites. Sheep are prone to attack from various parasites, such as ticks, mites, keds and blow-fly larvae. A badly infected animal suffers from itchy skin which can make it feel very ill and lose its appetite.

Each year shepherds protect their flocks by 'dipping', i.e. plunging the animals into a bath full of insecticide so that it soaks the coat of wool and kills the insects on the skin. In the early days of dipping, the insecticide would be an organo-chlorine-based compound, such as Lindane, but these were phased out by the 1980s in favour of organo-phosphorus (OP) compounds. However, even these are now falling out of use because it is alleged that farmers exposed to OP pesticides suffer from side-effects. (The same products have also been implicated in causing the so-called Gulf War Syndrome because the forces serving in the war against Iraq were protected from sandflies and other biting insects by spraying tents and bedding with OP insecticides.) The symptoms range from exhaustion and memory loss, to muscle weakness and depression. Farmers are now switching to sheep dips based on synthetic pyrethroids but extra care has to be taken in their use if the dip bath is situated close to a river or stream because of their extra toxicity for aquatic life. In 1995 and 1996, in the UK, there were some serious pollution incidents caused by the use of these new pesticides. For example, in July 1995, in the River Teviot in the Scottish Borders, about a thousand fish were killed by sheep dip containing pyrethroids. Similarly, in April 1996, almost all the mayfly larvae and shrimps in a 30 km stretch of the River Caldew in Cumbria were wiped out, whilst in Perthshire, the invertebrates in 25 km of the River Earn were killed by the same type of pesticide. The environment agencies are now contacting farming organizations to warn them of the environmental dangers of the new sheep dip compounds.

Organic waste from farming

The large numbers of livestock on Britain's farms all produce waste. Some livestock, such as poultry and pigs, are in most cases kept indoors all of their lives, i.e. they are raised intensively. The waste products have to be removed from the rearing buildings and disposed of, usually by spreading on the land. Winter weather is unsuitable for cattle to spend their days out in the fields; there is also a lack of grass for grazing and the weight of the animals walking over the ground in wet weather quickly turns fields into mud. For these reasons, the animals are brought indoors in the late autumn and don't return to the fields until the grass starts to grow again in the spring. As with poultry and pigs, the farmer has to dispose of the waste from the cattle for all the months they are confined to their stalls.

The spreading of animal manure onto fields can be carried out only in favourable weather conditions. This is usually on frosty days because the ground is frozen and can withstand the weight of the tractor and muck-spreader. In a mild damp winter, farmers have a frustrating time because they cannot get the waste spread and instead have to store it in large tanks or middens until the ground is firm enough to dispose of it. Care has to be taken when spreading animal waste onto the ground so as to avoid it entering water courses. We learned in Chapter 1, in the section on sewage treatment, that the amount of organic matter is measured as biochemical oxygen demand (BOD) and that untreated sewage has a BOD of about 300 mg/l. Animal slurry can have a BOD as high as 20,000! Even a small amount of waste entering a stream or being washed in by unexpected rainfall can have a disastrous effect on the water. The dissolved oxygen in the water (which may be only about 10 mg/l) is rapidly depleted and aquatic life suffocated.

A particular problem arising from cattle farming is that caused by silage liquor. All the time that cattle are indoors they have to be fed. They receive specially prepared cattle cake, hay, maybe some turnips, and silage. Silage is degraded grass and is prepared in the late spring and summer by cutting grass and allowing it to wilt for only a day or two. It is then collected with special equipment as shown in Cover Illustration 3 and put into a silo or a silage pit.

The grass is compressed by the tractor in the silage pit or by its own weight in the silo, and then covered with a waterproof membrane so that it degrades in the absence of oxygen. After a few months, the grass is converted into a moist solid which has the appearance rather like tobacco. During the degradation process, juices are squeezed out of the grass and

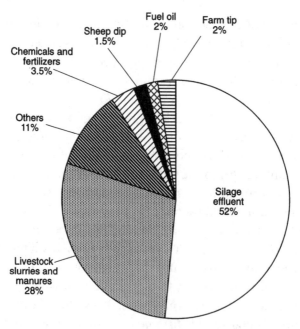

Figure 16. Water pollution caused by different agricultural activities in Scotland, 1982–90
Source: Scottish Agricultural Pollution Group, Pollution Review, 11, August 1994, Scottish Environment Protection Agency, Stirling

emerge from the silo or silage pit. This is called silage liquor; it has a very high BOD value, typically about 50–60,000. If this liquid accidentally gets into a stream it has an even worse effect than slurry. Of all the water pollution events that occur in the UK, probably the most serious, because of the numbers of fish killed, are caused by silage liquor. Figure 16 shows the proportion of different types of water pollution caused by agriculture in Scotland.

Notes

1.   Quentin Seddon, *The Silent Revolution. Farming and the Countryside into the 21st Century*, BBC Books, London, 1989.
2.   N.F. Gray, *Drinking Water Quality. Problems and Solutions*, Wiley, Chichester, 1994.
3.   G.A. Best and A.D. Ruthven, *Pesticides – Developments, Impacts and Controls*, Royal Society of Chemistry, London, 1995.
4.   *The Indirect Effect of Pesticides on Birds*, Report by UK Joint Nature Conservation Committee, Natural History Book Service, Totnes, Devon.
5.   C.H. Edwards (ed.), *Environmental Pollution by Pesticides*, Plenum Press, London, 1973.

Further reading

Agriculture and Pollution, Royal Commission on Environmental Pollution, 7th Report, HMSO, London, 1979.
Sustainable Use of Soil, Royal Commission on Environmental Pollution, 19th Report, HMSO, London, 1996.
Agriculture and the Environment, Proceedings of ITE symposium no. 13, March 1984, Institute of Terrestrial Ecology, Cambridge.
Prevention of Environmental Pollution from Agricultural Activities Code of Good Practice, Scottish Office, Agriculture, Environment and Fisheries Department, Edinburgh, 1997.

5. Fish farming

One of the most rapidly growing industries in the UK today is that of fish farming (also called aquaculture, although this term also includes shellfish and crustacea farming). These fish farms are either raising trout or salmon, although experiments are being carried out into rearing other species such as halibut and turbot. The industry in the UK is dominated by salmon farming, which takes place mostly in Scotland. However, the current UK production of 83,000 tonnes each year is small by comparison with Norway where the annual production is 220,000 tonnes. An indication of the speed of the development of salmon fish farming in the UK is given in Table 13 and Figure 16. World-wide, the production of seafood

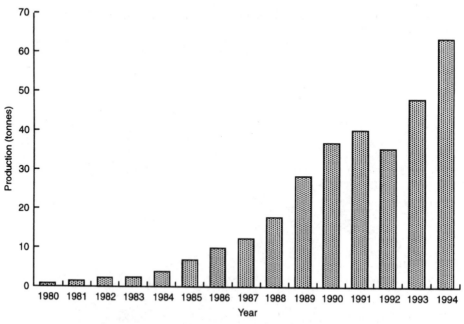

Figure 17. Growth of Atlantic salmon farming in Scotland, 1980–94
Crown copyright. Reproduced with the permission of the FRS Marine Laboratory, Aberdeen, Scotland

by aquaculture amounts to 115.9 million tonnes.[1] China dominates the market with an output of 27.3 million tonnes and much of the world-wide production is accounted for by shrimp farming (21 million tonnes).

Table 13. Annual UK production of salmon by aquaculture	
Year	Salmon produced (tonnes)
1982	2,000
1988	18,000
1991	52,000
1996	83,000
2000 (predicted)	132,000

Fish farming was probably started by the Chinese 4,000 years ago but in Britain there are early records of trout and carp being reared in monastery ponds to provide fish for the monks on meatless days, special feast days and for Lent. The first experiments in raising salmon in tanks started in the 1960s but it took more than ten years before it became a viable commercial venture.

In the UK, fish farming is largely confined to raising salmon and trout, and each fish has different requirements for successful rearing. Trout farming is a much smaller industry than salmon farming: in 1996, just over 4,000 tonnes were produced.

Salmon farming

The salmon farm attempts to reproduce the natural life cycle of the salmon which is shown in Figure 18, but accelerates it by the use of artificial feed, heat and lighting. The 'broodstock', i.e. the hen salmon full of eggs and the cock salmon which will be used to fertilize them, are moved from their sea cages to freshwater. The eggs are stripped from the female and subsequently fertilized by the male's milt. The fertilized eggs are kept in trays over which flows high purity water. When hatched, the fry are kept in freshwater tanks and fed an artificial diet.

After about 18 months, the young fish become smolts and are then transferred to cages moored in sea lochs. Again, they are fed with a specialized diet and grow rapidly to marketable size after about two years at sea.

The sea cages are normally circular or rectangular structures from which are suspended strong nets to a depth of 10–15 metres (Cover

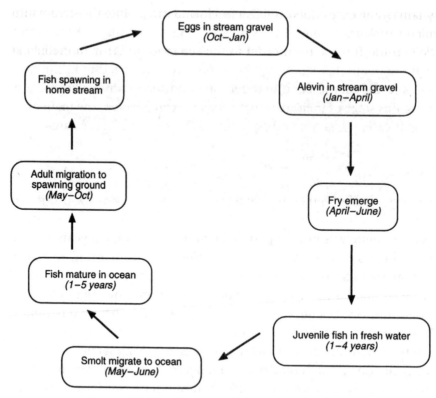

Figure 18. The life cycle of the Atlantic salmon

Illustration 4). These cages should be anchored to the sea bed in such a position so that there is a good flow of clean sea water through the nets. However, if they are sited in too exposed a position, they can be damaged by storms. The fish farmer usually has to find a compromise on these two requirements and the great majority of units are moored in sheltered lochs.

A recently designed cage unit is capable of withstanding hurricane-force winds and these can be sited further away from the coast. There are considerable benefits for the marine environment from this innovation.

In the initial stages of production, the newly hatched fry are kept in indoor tanks through which flows high-purity water from a nearby stream. When they have grown to the parr stage they are moved to larger tanks, either indoors or outside, where they are given a special feed which is high in protein and fat to speed growth. The feed also contains other minor but important additives such as vitamins, fatty acids, amino acids, phosphorus, zinc, manganese and a pigment. The overflow from the tanks contains some excess food and fish faeces and can pollute the stream into which it empties.

When the fish are transferred to the sea cages to grow to their final weight of about 2–5 kg, many thousands are kept together in the cage units and they may spend between one and three winters at sea. It is during this stage of their lives that problems can develop, the main ones being disease, predation and pollution of the surrounding sea area.

Disease

Crowded together in their tens of thousands, salmon are very susceptible to disease, and this can spread quickly through the stock resulting in heavy losses for the fish farmer. The three main problems are pancreas disease which is caused by virus and is untreatable, sea lice parasites, and the bacterial disease furunculosis.

Sea lice are a particular scourge of the fish farmer. They begin their life as small, free-swimming, flattened insects about 3 mm long, which attach themselves to the tail of the salmon by biting into the skin and then steadily move towards the head, growing and maturing all the time. At the adult stage they become firmly attached to the head, which they then gradually eat into, forming large open wounds on the fish. The salmon are obviously very distressed by these infestations (a fish with a bad infestation may have more than ten lice on its head), and have depressed appetites as well as being susceptible to infections of the wounds. The fish farmer needs to remove the parasites because the fish will not grow fast enough for marketing.

The long-established technique for eradicating fish lice is to treat the infected cage with a pesticide. This kills the parasite but has minimal effect on the fish. The fish cage is enclosed in a tarpaulin and a pesticide called dichlorvos is added so that the final concentration in the enclosed water is about 1 mg/l. After sufficient time to kill the parasites, the tarpaulin is removed and the water is flushed into the surrounding sea. Dichlorvos is an organo-phosphorus insecticide (see Chapter 4) and has also been used as a fly spray and in sheep dips. The manufacturers claim that it is biodegradable and that, once released into the surrounding sea, it quickly becomes harmless. Other people though, such as anglers and those who collect crabs and lobsters for sale, claim that dichlorvos is very poisonous, particularly to the larval stages of crustacea. There has been a long battle between the two factions and fish farmers are now looking to alternative means of eliminating the lice. At some farms they use hydrogen peroxide; this is readily broken down once used, but much larger amounts are need to kill the parasites. More recently, investigations have been carried out on other substances, either another pesticide or an 'in feed' treatment. The alternative pesticide is a synthetic pyrethroid (see page 5) called cypermethrin, which is regarded as less hazardous to the environment because it is based on a natural product. The other approach is to incorporate a chemical in the fish feed. This has the advantage that it doesn't so readily enter the surrounding sea (though traces may be present in the fish faeces), but it does become incorporated in the blood of the fish. When the lice bite into the salmon flesh they suck in some of the chemical and this then kills them.

An interesting new development in sea lice control is to use a 'cleaner' fish. It has been found that a species of fish called a wrasse has a healthy appetite for sea lice! If a small number of these fish are put into the same cage as the affected salmon, they remove the lice from the salmon. This is beneficial for the salmon because its parasite is removed, for the wrasse because it has eaten well, and for the fish farmers because their fish stock are healthier and there has been no harm to the environment.

Predators

There are many animals and birds that feed on fish and some of them become a nuisance to fish farmers. Of particular concern are seals, otters, herons, cormorants and other sea birds.

The birds can be deterred from catching the young and growing fish at the fish farm by netting the surface of tanks and cages, but this has to be done carefully to ensure that the birds don't get caught in the netting. All the wild birds that prey on farmed fish are protected by law (Wildlife and Countryside Act 1981) so fish farmers are not allowed to trap or shoot the predators. The same applies to the mammals. Seals can be a particular pest because they try to enter the cages under water and can damage the nets. This can result in the farmed salmon escaping if it is not noticed.

A variety of techniques are used to deter seals from interfering with the netting cages. The most commonly used technique is to suspend an extra net outside the fish cages and this prevents the seals getting access to the stock. In some farms they use sonic scarers – devices which emit that objectionable noise to the seals which, frightens them away. Some farmers have resorted to shooting persistent animals, although this is against the law (Conservation of Seals (Common) Order 1988). 206 seals were reported to have been shot in one year in the late 1980s.[2]

Pollution

Salmon fish farms can have a major impact on the marine environment. Most sea cages are moored in sea lochs and inlets that previously received minimal waste from the sparse population in the surrounding area. With the arrival of a fish farm there is a considerable increase in the amount of waste entering the water. Figure 19 summarizes the main sources of these wastes.

Fish are fed artificially from the surface with pelleted feed which slowly sinks through the cage. The feeding rate is designed to minimize wastage because fish feed is expensive. Nevertheless, excess, uneaten feed falls through the water and is deposited on the sea bed. In addition, there is a constant 'rain' of faeces from the fish which contains organic matter as well as traces of minerals and nutrients. The fish also excrete urine so there is an input of soluble waste into the water.

The main concerns are about the accumulation of organic matter on the sea bed, and the addition of extra amounts of nitrogen and phosphorus to the water. There are also unproven fears that the residues of antibiotics and other treatment chemicals may harm other marine life in the surrounding sea area.

The organic matter settles as a thin layer on the sea bed where it smothers the sand, stones, rocks or mud. If the accumulation is severe, then the decomposition of the organic matter by naturally occurring bacteria results in a depletion of the dissolved oxygen in the water within the sediment. This is the same effect as described in Chapter 2 when a poor quality sewage effluent enters a river. Sometimes, the growth of bacteria is so vigorous that thick white mats of them develop over the waste. The bacteria consume the organic matter and use up oxygen dissolved in the water at a greater rate than it can be replaced by clean water. This can give rise to anoxic conditions (no oxygen) in the sediment and result in the formation of harmful and objectionable by-products. In the worst cases, hydrogen sulphide, ammonia and methane gases bubble up from the sea bed to the surface. Hydrogen sulphide is a poisonous gas and can kill the fish which are the originators of the waste: a good example of 'fouling one's own nest'. These bad conditions are rare because there is usually sufficient movement of water over

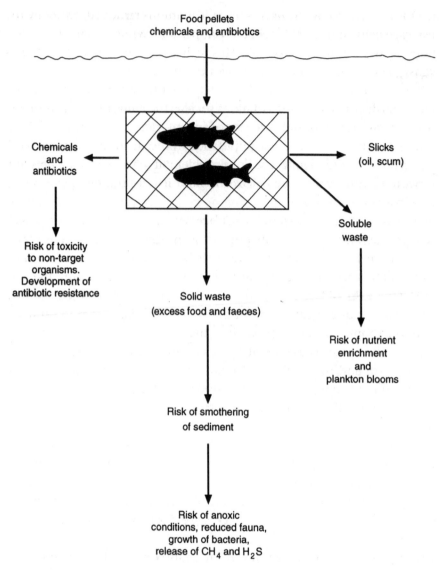

Figure 19. Environmental impacts of cage fish farming on the marine environment

the sea bed to ensure enough oxygen is present for the decomposition of the waste. A more common effect of the excess organic matter is to encourage a build-up of a population of marine organisms which thrive on the waste. The uneaten food and faeces provide a valuable food source for marine worms such as the polychaetes *Capitella capitata* and large numbers accumulate below fish cages.

Once the source of the excess organic matter is removed, either by the fish cage being moved or it being left empty for a while (to fallow), conditions slowly return to normal. This is because the organic matter is consumed or broken down by the worms and bacteria until there is none left and the organisms starve. This recovery can take between two and ten years depending on many factors such as the amount of accumulated waste, water temperature, water movement, etc.

The phosphorus in the waste is usually present in its insoluble form and is not considered to be a serious problem in sea water. It is a problem though in fresh water (see below) because it is a limiting nutrient (see page 30). In sea water, the limiting nutrient is nitrogen, so extra inputs of this can encourage the growth of algae (phytoplankton). There is a possibility that the soluble nitrogen waste produced by fish in a fish farm could result in the formation of algal blooms. However, there is no clear evidence that this has occurred despite a number of investigations.

The doubling time for marine phytoplankton in favourable conditions is about two days but for this to happen, there has to be sufficient sunlight, nutrients and calm water. The combination of these factors does not happen often because most fish farms in the UK are located on the West coast of Scotland which is not renowned for calm sunny weather. Algal blooms therefore are rare events. They have occurred though and have had disastrous effects because some marine algae produce toxins in the water as a result of their photosynthesis. Blooms of toxic algae have been responsible for mortalities of farmed and wild fish in Scotland, Norway and Ireland.[3]

Trout farming

The artificial rearing of trout has been taking place in the UK and elsewhere for much longer than salmon farming and it is much more widespread. Trout rearing not only produces fish for fishmongers and supermarkets, but also maintains or improves the fish stocks in rivers for anglers.

As with salmon farming, the eggs are stripped from a mature female, fertilized by milt from a male and then hatched in a hatchery. After spending their juvenile stages in large tanks through which clean river water circulates, the young fish are then transferred to large ponds or to cages suspended in a freshwater lake or loch.

The same environmental problems arise from trout farms as with salmon farms. Waste food and faeces are present in the outflow water from fish-rearing ponds and these can give rise to the same effects on water quality as those of a sewage discharge, namely, the dominance of pollution-tolerant organisms in the aquatic life, depletion of dissolved oxygen and the increase in organic matter and nutrients. Artificially raised trout are prone to a variety of diseases which have to be controlled by treatment chemicals. The most common diseases are caused by bacteria or by a fungus and these are treated by adding formalin or Malachite Green. Fortunately trout are not so seriously affected by parasites and are not attacked by lice, so fewer chemicals are used in trout farming than salmon farming.

The main problem associated with trout farming is the release into the surrounding water of nutrients from excess feed and waste from the fish. As mentioned earlier, the input of phosphorus in particular can activate the development of algal blooms because this nutrient is usually in short supply for algal growth in unpolluted fresh water.

Loch Fad trout farm, Isle of Bute

Loch Fad is a freshwater loch, 2.4 km long and 0.3 km wide, situated just inland from the seaside resort of Rothesay on the Isle of Bute, Scotland. It has been used for trout farming since 1976 and, at one time, was the largest trout farm in Europe, producing over 300 tonnes of fish each year.

The combination of waste feed and faeces from the large number of fish confined in this small loch has had a marked effect on the water quality. The sediment is very enriched with organic matter and this has encouraged the growth of a huge population of tubificid worms which feed on waste. In some areas close to the cages, the worms are at a density of more than $20,000/m^2$ compared with about $10/m^2$ in a clean water loch.

The most noticeable effect of this intensive aquaculture, though, is the development of algal blooms in summer months because of the presence of high concentrations of phosphorus. This nutrient chemical is discharged into the loch in the waste from the fish and from the decomposing waste feed. In 1989–92, the concentration of total phosphorus exceeded 100 µg/l which, according to the classification of waters for eutrophic status (see Table 7), puts Loch Fad into the hypertrophic class.

The algal blooms on fine summer days in the early 1990s were so intense that the outlet stream had the appearance of green paint and the sea in Rothesay Bay developed a distinct green colour. Not surprisingly, the fish in the cages suffered from the algal blooms and there were two main effects. Firstly, some of the algae which developed were of the blue-green variety which produce a toxin in the water as they photosynthesize. Secondly, the photosynthesis of the huge number of algae caused marked fluctuations in the levels of dissolved oxygen in the loch. During the day, the photosynthesis produced excessive amounts of oxygen so that the water became supersaturated (over 120 per cent). At night, the algae respired and used up dissolved oxygen, and the levels decreased to 30 per cent saturation (about 3 mg/l which is below the danger level of 5 mg/l).

After this time of peak production, the farm output declined because of the poor water quality: by 1993, the amount of fish in the cages was only 42 tonnes. However, the water quality improved because of this reduction in fish numbers. In the period 1993–95, the concentration of total phosphorus had declined to 75 µg/l which classified the water as being eutrophic. By 1996 the production had increased to 100 tonnes/year.

Algal blooms continued to be a problem even though the concentration of nutrients had declined. To counteract them, the loch was treated with barley straw (see page 37) in March 1993 and in February 1995, and this seems to have been effective as the algae appear less abundant than in earlier years.

Notes

1. *Fish Farm International*, Vol. 24, No. 10, October 1997.
2. *Fishfarming and the Safeguard of the Natural Marine Environment of Scotland*, Nature Conservancy Council, Edinburgh, 1989.
3. ICES Special Meeting on the Causes, Dynamics and Effects of Exceptional Marine Blooms and Related Events. Copenhagen, October 1984. International Council for the Exploration of the Sea.

Further reading

Fish Farming and the Scottish Freshwater Environment, a report prepared for the Nature Conservancy Council by the Institute of Aquaculture, Institute of Freshwater Ecology and the Institute of Terrestrial Ecology, NCC Contract No HF3-03-450, June 1990.

6. Tip drainage

Earlier in this book we dealt with the pollution problems that arise from the disposal of our liquid waste. We also get rid of an enormous amount of solid waste in the form of the refuse that is put into our bin sacks or 'wheelie' bins. It's difficult to visualize the amount of rubbish we throw out but Cover Illustration 5 shows the accumulated waste that a 'typical' family would put into their rubbish bin in one year.

Despite government attempts to get us to recycle our refuse, the great majority is collected by the local authorities and disposed of. Some is incinerated and the heat produced is occasionally used to heat large buildings or to produce steam for generating electricity. Some is converted to a compost for improving the quality of municipal land. Most however is dumped at landfill sites, areas of derelict or poor quality land, worked-out quarries or open-cast coal mines which are filled with waste to convert them into more usable land. Figure 20 shows the relative proportion of waste disposed of by various methods.

If you think about what is put into your waste bin at home, you will notice that it consists of plastics, shopping bags, wrappings, paper (news-

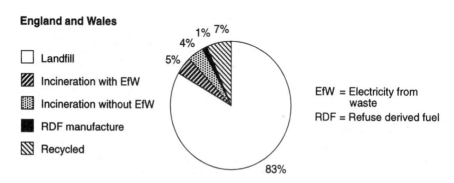

England and Wales

☐ Landfill
▨ Incineration with EfW
▦ Incineration without EfW
■ RDF manufacture
▨ Recycled

1% 7%
4%
5%

EfW = Electricity from waste
RDF = Refuse derived fuel

83%

Total reported = 22.41 million tonnes

Figure 20. Relative proportion of disposal methods for solid waste
Source: The Environment of England and Wales, *a snapshot, Bristol, 1996*

paper and packaging), metal (mild steel and aluminium cans), organic matter (waste food, garden waste), and glass from jars and bottles. The quantities of these various types of domestic rubbish have been measured by government scientists. They have found that the relative amounts of solid refuse by weight are as shown in Table 14.[1]

Table 14. Composition of domestic waste, 1992

Type of waste	Percentage by weight
Plastic film	5
Plastic bottles and packaging	3.5
Glass	10
Rags	4
Ferrous metals	7
Aluminium cans and foil	1
Paper	32.5
Waste food	20
Dust, etc.	10
Miscellaneous items	7

This analysis shows only the weight of the refuse. When you consider that plastics are much lighter than, say, glass, then you can see that plastic waste will occupy a much greater volume of the rubbish.

The composition of our waste today is very different to that of 100 years ago. Plastics hadn't been invented then and most people had coal fires in their homes, from which the ashes were put into the dustbins. The composition of waste from 1890 was as shown in Table 15.

Table 15. Composition of domestic waste, 1890

Type of waste	Percentage by weight
Cinders and ash	64
Fine dust	20
Vegetable & animal waste	4.6
Waste paper	4.3
Straw	3.2
Glass	1.4
Metals	1
Crockery	0.6
Bones	0.5
Rags	0.4

They say you can learn a lot from people's waste and a comparison of these two lists clearly demonstrates how much more food we waste and wrappings we throw away.

At the landfill site (or waste tip) the rubbish brought in by the collection vehicles is first dumped, then spread out evenly and compacted by specially designed vehicles (Cover Illustration 6). Ideally, at the end of each working day, the refuse should be covered with top soil to prevent it being infested with flies, rats and seagulls. Eventually the tip site is completed and is covered with soil and planted with grass. It can then be used for farmland or even for golf courses and play parks.

Throughout the time of the formation of the landfill site, and for many years after its completion, nearby streams and rivers can be polluted by drainage water from it. This drainage water is called leachate and originates from rainfall or spring water which percolates through the decomposing waste. As it does so, it picks up pollutants from the waste. The most common constituents are organic matter from the rotting waste food and garden refuse, and iron from the rusting cans and other pieces of mild steel such as old bicycles, etc. The leachate can also contain high concentrations of inorganic ions such as chloride or toxic metals. Table 16 shows the typical results of the chemical analysis of a sample of leachate from a waste tip.

Table 16. Chemical analysis of leachate from a new waste tip

Chemical determinand	Concentration (mg/l except pH)
Biochemical oxygen demand (BOD)	830
pH	7.5
Ammonia nitrogen	120
Oxidized nitrogen	<0.5
Chloride	1214
Iron	380

This sample was obtained from a tip site where the rubbish had been recently tipped and these results show that the leachate was very polluting. The waste was decomposing rapidly and there was sufficient rain water to wash the pollutants from the site.

The analytical results vary greatly from different tip sites and also from

Table 17. Chemical analysis of leachate from an old tip site

Chemical determinand	Concentration (mg/l except pH)
BOD	4.0
pH	6.8
Ammonia nitrogen	5.0
Oxidized nitrogen	4.9
Chloride	367
Iron	0.2

the same tip site over time. Eventually, the rate of decomposition slows down and the tip site stabilizes. The results shown in Table 17 are typical of drainage water from a tip site that has been in operation for many years.

Once the leachate reaches a stream or river, the organic matter is broken down by naturally occurring bacteria on the river bed (biodegradation) in the same manner as described earlier for treated sewage effluent (Chapter 2). If the concentration of organic matter is high, and there is little dilution in the receiving stream, the dissolved oxygen levels will be reduced and this can have an adverse effect on the aquatic life. The iron in the leachate is usually present as insoluble ferric hydroxide and appears as fine particles of rust. These particles give the stream a distinctly orange colour. They slowly settle to the stream bed so that, with increasing distance from the source of the leachate, the stream gradually becomes clearer. The accumulated iron has an adverse effect on the stream because, not only is it discoloured, but the particles of rust coat the leaves of water weeds, reducing their photosynthesis. They also clog the gills of insect larvae and fish so they swim away or die because they cannot extract oxygen from the water.

In order to stop this type of pollution occurring, water must be prevented from entering the decomposing waste as much as possible. This is done by constructing drainage channels to collect water and direct it away from the tipping area. If water does get into the rubbish and later emerges as polluted leachate, it should be collected and purified before it enters a stream. The organic matter in the leachate can be treated in the same way as described for sewage and at some tip sites the operators build a small treatment works.

If the concentration of organic matter is relatively small but the leachate contains a great deal of iron, then the treatment method is usually to precipitate the iron and allow it to settle in large shallow settling ponds. The rate of precipitation can be increased by adding lime or special chemicals called coagulants, such as polyelectrolytes.

Note

1. *Recycling Waste Management* Paper 28 Department of the Environment, 1992.

Further reading

Making Waste Work: A Strategy for Sustainable Waste Management in England and Wales, Cm 3040, HMSO, London, 1996.

7. Mine-water pollution

In the UK, about 200 km of river lengths are polluted by mine water and a further 400 km by discharges from abandoned metal mines. The problem is likely to get worse because the closure of many deep-shaft coal mines in the early 1990s will result in further flooding and discharges of polluted mine water. In a survey of pollution caused by mine-water drainage in Scotland, it was found that there were 59 discharges of mine water: 25 from abandoned mines, 17 indirectly associated with mining, 8 from active mines and the remainder from coal waste tips and open-cast mines.[1]

When a coal mine is in use, the tunnels and shafts are cleared by pumping out the water which seeps into them. When the coal has all been extracted, the mine is closed and the pumping equipment is removed. Gradually the mine workings flood with ground water until eventually it reaches the surface and flows into a stream. This process can sometimes take many years after closure depending on the depth of the mine shafts, the rate of flooding and the extent of the workings.

In some areas the ground water that emerges from abandoned coal mines is of good quality and can be used for public supply, but these instances are rare. Usually, the mine water is polluting because of the high concentration of various ions, particularly sulphate and iron (water containing high concentrations of the latter is called ferruginous). This is because, adjacent to coal seams, the ground contains pyrites, FeS_2. With the opening of the mine and the passage of air through the tunnels, the pyrites becomes oxidized to ferric sulphate $Fe_2(SO_4)_3$. The closure and subsequent flooding of the mine cuts off the oxygen and the ferric sulphate is reduced to ferrous sulphate, $FeSO_4$. This reduced form of iron is water soluble so the flooding mine water becomes contaminated with iron. The ground water also becomes acidified because of the presence of sulphur bacteria in the walls of tunnels. These bacteria utilize the oxygen present in iron sulphates as an energy source because of the lack of oxygen in the air and, as a result, sulphuric acid is formed. These chemical reactions can be summarized as follows:

$$2FeS_2 + 7O_2 + 2H_2O \rightarrow 2(FeSO_4) + 2H_2SO_4$$
$$2FeSO_4 + H_2SO_4 + {}^1/_2O_2 \rightarrow Fe_2(SO_4)_3 + H_2O$$
$$Fe_2(SO_4)_3 + 6H_2O \rightarrow 2Fe(OH)_3 + 3H_2SO_4$$

The result of these reactions is that the mine water, as it emerges from the ground some time after abandonment of the mine, has the following characteristics:

- it is devoid of oxygen
- it has high levels of sulphate ion
- it contains soluble iron in its reduced, ferrous, form
- it is acidic.

Dalquharran mine-water pollution: a case study[2]

The Dalquharran mine is situated near the village of Dailly in South Ayrshire, Scotland. Coal has been extracted in this area since the fourteenth century, initially from the surface because the layers of coal emerged from the hillside. As time went by, successively deeper layers were mined until there was a honeycomb of workings throughout the area.

In the 1950s, the then National Coal Board (NCB) established the Dalquharran Mine and coal was extracted from six different layers at increasingly deeper levels. The workings were subject to flooding so pumping equipment was installed and the mine water was discharged into the nearby River Girvan. This river was of very high quality and was popular with anglers because of the many trout and salmon that could be caught in it. Unfortunately, the mine water was polluting because it contained coal solids as well as iron from the oxidation of the pyrites which were also present in the coal seams.

The NCB received many warnings from the local River Purification Board (the predecessor regulatory authority to the Scottish Environment

Protection Agency). Eventually it was prosecuted in 1975 and fined £20 for causing pollution! This did not prevent the pollution, however, and the Coal Board was threatened again with prosecution. By now, the coal had mostly been extracted and plans were being made to close the mine. This took place in May 1977 and the pumps were removed from the tunnels. The mine slowly started to flood and, from the rate at which the water rose, it was predicted that it would eventually come to the surface by the end of 1979.

The outbreak occurred in October 1979 and the effect was spectacular. It was always anticipated that the mine water would contain iron and would be acidic, but the concentrations were much higher than expected. The chemical analysis of the mine water when it first emerged was as follows:

pH	4.0
Conductivity (µS/cm)	4259
Sulphate (mg/l)	5490
Iron (mg/l)	1093
Aluminium (mg/l)	92.0

The flow rate of the discharge was 227m^3 per day and the effect on the river was disastrous. For the 16 km length of river, from the discharge point to the sea at the Firth of Clyde, the water was stained bright orange. The river bed rapidly became coated with insoluble ferric hydroxide and all the fish and most of the invertebrate life were killed. All that survived were those macro-invertebrates that could tolerate the pollution. Cover Illustration 7 shows the condition of the river soon after the outbreak occurred.

The pollution in the river was so bad that even the sea was discoloured as the iron was washed into it by the river. Fishing boats in the harbour were stained by the water and there were many complaints from holiday makers who visited the attractive seaside town of Girvan.

The National Coal Board was taken to court for causing the pollution but the sheriff at Ayr found the company not guilty! His view was that the NCB had merely closed the mine and was not responsible for the subsequent flooding, because it would have taken place naturally. However the case was taken to a higher court in Edinburgh for an appeal against this decision and this time the NCB was found guilty. The Appeal

Court judges' opinion was that mining had caused the pollution. If mining hadn't taken place, the water would not have become polluted and come out of the ground. They said that the very act of starting the mine set in train a series of events that would eventually result in mine water entering the River Girvan.

As a result of this judgment, the NCB had to do something about the pollution. The first suggestions were to construct a special pipeline and pump the mine water to sea after some treatment, or to build a treatment plant adjacent to the mine and precipitate the iron into special large settling tanks. Both of these plans were too expensive for the NCB and other ideas were formulated:

1 One of the reasons for the outbreak was that water was flowing into the mine from surface springs and streams, so these could be sealed or diverted.
2 The acidity of the mine water could be neutralized by pumping lime into the workings and it was hoped that this would also precipitate the iron in the tunnels and shafts.
3 Investigations showed that the most polluted water was coming from the deepest part of the mine, so a special tunnel could be drilled into the surface layer of the mine to intercept the water before it flowed to the lower layers.
4 Finally, it was proposed to trap any mine water in a specially constructed holding tank. From here it could be released into the river only when the river flow was high after rain.

These proposals were accepted and the work completed in late 1983. In addition to all this work, the Coal Board made a donation of £1,000 to the local Fisheries Board to restock the river with salmon fry. Meanwhile, the quality of the mine water was slowly improving after its initial poor state and Table 18 shows the change in some of the chemical measurements with time.

The NCB's scheme was successful and, five years after the initial outbreak, the river was largely restored to its original state. Salmon returned and the iron on the river bed was slowly washed out to sea by the high winter flows.

One problem remained, however. In long dry spells in summer, the river flow decreased, but the holding tank steadily filled up with mine

Table 18. Chemical quality of the mine water from Dalquharran Mine (mg/l except for pH and conductivity)

	1980	1981	1994
pH	4.3	5.3	6.0
Conductivity	4,583	3,506	3,477
Sulphate	5,145	3,405	1,660
Iron	1,044	600	193
Aluminium	32	5.6	4.0

water and eventually it overflowed to the river. The pollution problem recurred (though it was by no means as bad as the effect in 1979), so a solution had to be found to prevent this happening in dry weather. The technique used was to create reed beds and make use of natural filtration from the soil and the roots of the plants (Cover Illustation 8). Reed beds are increasingly being used to purify wastewaters and to give an extra stage of treatment to sewage effluent, particularly if the effluent flows into a sensitive river.

For a reed bed to be successful the ground has to be levelled and then a drainage channel created which allows the wastewater (or mine water in the case of Dalquharran) to meander over the ground and pass through the soil layers. The area is planted out with various reeds which soon get established in the saturated ground. As the polluted water from the holding tank passes through the reed beds, the iron settles on the soil or is trapped in the reed's root network. By the time the effluent flows into the river it is purified and causes no problems to the River Girvan.

The incident at Dalquharran is by no means unique and there are many other abandoned mines which are polluting nearby rivers with ferruginous ground water. One of the worst cases in England was from the Wheal Jane mine in Cornwall.[3] Here, as in Scotland, the mine water is now being purified by various techniques including the use of a reed bed. It has been predicted that new discharges will occur in the future, especially in the north-east of England in County Durham because of the closure of a number of deep coal mines there.

Notes

1. *Abandoned Mines and the Water Environment*, Report of the National Rivers Authority, Water Quality Series No 14, HMSO, London, 1994.

2. D. Hammerton, 'Acid minewater pollution from an abandoned mine', in Best, G.A., Bogacka, T. and Niemirycz, T. (eds), *International River Water Quality*, E. & F.N. Spon, London, 1997.

3. Hamilton, R.M., Taberham, J., Waite, R.R.J., Cambridge, M., Coulton, R.H. and Hallewell, M.P. 'The development of a temporary treatment solution for the acid mine water discharge at Wheal Jane', 5th International Minewater Congress, Nottingham, 1994.

Further reading

Commission on Energy and the Environment, *Coal and the Environment*, HMSO, London, 1981.

Hester, R.E. and Harrison, R.M. (eds), *Mining and Its Environmental Impact*, Royal Society of Chemistry, London, 1994.

8. Acid rain

Acidification of rain and snow may seem to be a recent environmental pollution problem. However, the phenomenon has been known for over a century since it was first noticed that buildings, trees and plants were damaged if they were downwind of chemical factories discharging acid fumes. The damage was mostly confined to periods of rainfall because of the removal of air-borne pollutants by rain droplets. At that time, the problem was a local one, confined to an area close to the factories. This was because the factory chimneys were relatively short and there was not widespread dispersion of the pollutants. The air was also polluted with the smoke from the individual coal fires in all the houses, as well as from small power stations adjacent to the towns and cities that burned coal to produce electricity.

Nowadays there has been more centralization of heavy industry, and electricity generation takes place at fewer larger power stations which use a greater variety of fuels than, say, 50 years ago. The waste gases from these industries and power stations are usually discharged into the atmosphere from high stacks (cover illustration 9) so the gases are dispersed much more widely.

The main acidifying gases are sulphur dioxide (SO_2) and various oxides of nitrogen such as nitrous oxide (N_2O), nitric oxide (NO) and nitrogen dioxide (NO_2). These are collectively referred to as NOx. The SO_2 originates mostly from power stations whilst road traffic is the main source of NOx. These gases undergo a series of chemical reactions with cloud water and sunlight to form sulphuric and nitric acids, as shown in Figure 21.

As a result of using high chimneys to prevent the fumes from factories and power stations affecting the local population, the polluting gases can sometimes cross national boundaries and get washed down by rain into a different country from that in which they originated. For example, the Norwegians and Swedes have shown that they receive ten times more acidity in rainfall from other countries than they produce themselves.[1]

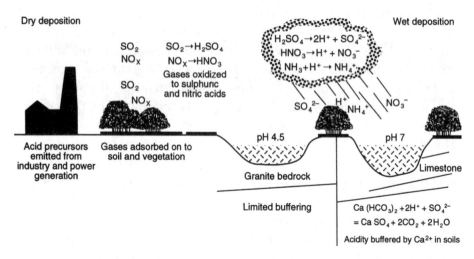

Figure 21. Chemical reactions of polluting gases in the atmosphere
Source: The Freshwaters of Scotland *(eds Maitland, Boon and McLuskey).*
Copyright John Wiley and Sons Ltd. Reproduced with permission

Much of this acidity originates from industrial areas in the UK. Similarly, Canada is a net importer of polluted air from the Midwest states of the USA and Japan receives acid rain from China.

The amount of acidity is expressed in terms of pH units on a scale of 0–14 where pH 7 is neutral. Values greater than 7 are increasingly alkaline whilst values less than 7 are increasingly acid. The pH scale is logarithmic so that a reduction of pH from 7.0 to 6.0 is a ten-fold increase in acidity.

Rainfall is naturally slightly acidic (about pH 6.0) because it dissolves carbon dioxide from the air to form carbonic acid. If rainfall has a pH of less than 5.6 it is regarded as being polluted with acid gases. Very polluted rain has a pH value of 3.0 or less.

The pH of water is usually measured with a pH meter as described in Chapter 12, but these instruments are a relatively recent invention. However, the Scandinavians claimed at a United Nations conference in 1972 that their rainfall became acidified by pollutants from other countries as early as the 1950s and in turn acidified many of their lakes. They did not have good chemical evidence for this because of the unreliability of pH measurements from that period, but produced other interesting testimony from an examination of their lakes. There are an estimated 15,000 lakes in southern Sweden that have been acidified by acid rain in the past 30 years! The change in pH of lake water over a period of time can be obtained by careful examination of the sediment at the bottom of the lake.[2] The sediment contains the skeletal remains of tiny algae called diatoms which, when they were alive, lived by floating in the surface layers of the lake. There are many different types of diatoms and each of them has a different tolerance to the acidity in the water: at a pH of, say, 5.5, the water is suitable for one group of diatoms but if the pH falls to 4.7, it will be too acid for some of them but another group will thrive. The diatoms have a relatively short life, only a few days, and when dead their remains sink slowly to the bottom of the lake. Once there, they get covered with other debris over time. To obtain a record of the change in the acidity of a lake over a long time, scientists obtain cores of sediment from the lake by pushing a long tube into the mud. These cores are then frozen and sliced into segments. One part of the segment is examined carefully for the skeletal remains of the diatoms (Cover Illustration 10) and these are identified to see what type they are and their preferred pH. Another portion of the segment is 'dated' by looking at the isotopes present in it, e.g. by carbon dating. From this examination of the length of the core, any changes in pH over time can be obtained. As an example, Figure 22 shows the pH records of three lakes in different parts of the UK which have been acidified over time. These show that there was a rapid increase in the acidity (reduction in pH) in the period 1950–60 which

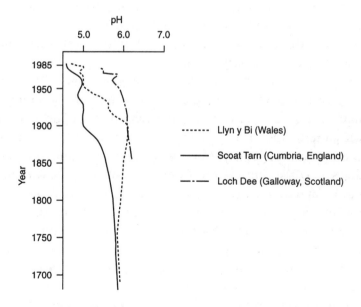

Figure 22. pH changes in three UK lakes according to the diatom records
Reproduced by courtesy of ENSIS, University College, London

coincided with the industrial expansion in western Europe and the construction of large power stations with their high stacks.

Although we have a number of lakes which have been acidified by acid rain in the geologically sensitive areas of the UK, the situation in Scandinavia is much more serious. In southern Norway, half the brown trout population has been lost by acidification in more than 2,800 lakes, whilst in Sweden, of the 85,000 lakes with a surface area greater than 2 hectares, 15,000 are acidified and, of these, 4,500 are fishless. There are two main reasons for this:

1. The prevalence of south-westerly winds which bring polluted rain from the UK, Germany and other countries in that direction.
2. The geology of the country which is mostly of granite, schists and quartzite which are slow weathering and resistant to erosion.

Although the phenomenon of acid rain has been known for some time in the Scandinavian countries and Canada, it has only recently been causing concern in the UK, especially in the northern part of Britain where acid soils and resistant rocks predominate. Figure 23 shows the areas in the UK where the surface water and ground water are susceptible to acidity because of the nature of the geology.

The prevailing wind in the UK is from the south-west which brings in uncontaminated rain from the Atlantic Ocean. Figure 24 shows a typical weather map with a depression centred off north-west Britain bringing in strong, wet south-westerly winds.

On a significant number of occasions, however, the weather is from the south-east because a depression can pass through the southern part of Britain. Figure 25 shows a weather map that produces this sort of weather pattern. The air movement round the depression draws in air from the near continent. Any rain that falls on Scotland is polluted from discharges into the atmosphere from the industrial areas around the English Midlands as well as from further south. An overall measurement of the acidity in rainfall in northern Britain shows that there is a gradient of acidity from high values (low pH) in the south-east and lower values in the north-west (Figure 26).

However when the geology of the area is considered (see Figure 23) the rocks in the south-east are seen to be softer and to have adequate neutralizing capacity to counter the effects of the acid rain. The rivers and streams in this area are not acidic. This is also a relatively dry area so the

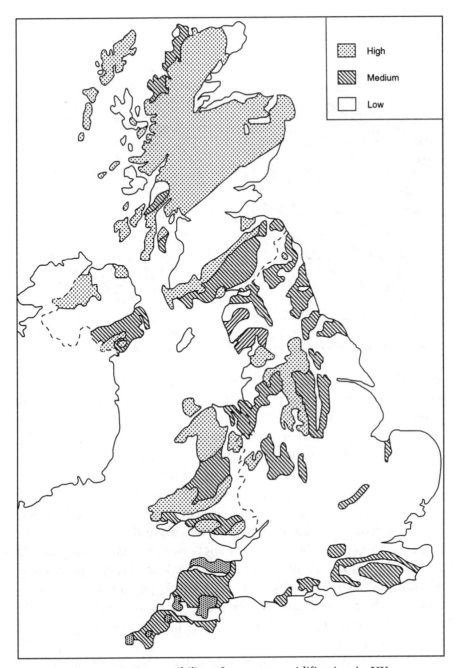

Figure 23. Susceptibility of waters to acidification in UK
Source: Journal of the Geological Society 143. *The Geological Society, Piccadilly, London.*
Reproduced with permission

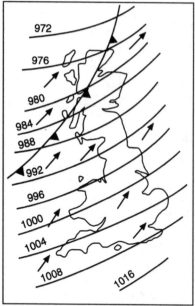
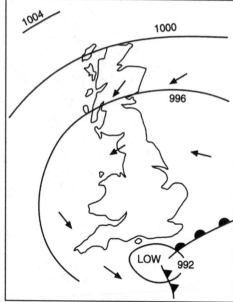

Figure 24. Weather map for south-west winds in the UK

Figure 25. Weather map for south-east winds in the UK

amount of rainfall is much less than in other parts of the UK. The problems are most pronounced in the Western Highlands, the Galloway Hills in south-west Scotland, the Lake District and parts of Wales. In all of these areas, the rainfall is higher than average so the loading of acidity (the amount of rainfall × the acidity) is greater. These areas also have resistant geology so any acidity is not neutralized before it enters the streams.

So what are the effects of acid rain on receiving streams? In Scandinavia, Canada and north-east USA, there is clear evidence which links acid rain to the death of fish and also making some lakes unfit for fish life.[3] It is a particular problem at the time of the snow melt. Sometimes the snow is acidified by acidic pollutants if the wind is in a certain direction; it also gets discoloured because of soot particles which get trapped in the snow crystals. When this so-called 'black snow' melts, the acid in it rapidly reduces the pH of the melt water and kills the fish and other sensitive aquatic organisms.

In Scotland, north-west England and Wales, the effects are not so dramatic, but surveys carried out by water scientists have shown that, for example, some lochs in the Galloway Hills which were at one time good trout fisheries but are now fishless.[4] The main problem with acid rain is not so much the acidity present (as H^+) in the water but the excessive

Figure 26. Gradation of rainfall pH across the UK

amount of sulphate ion (SO_4^{2-})which has been formed from the oxidation of SO_2. (The measurement of sulphate is described in Chapter 12.) As the sulphate ion passes through the soil and over the rocks, it needs an 'ion pair' to maintain the rules of chemistry of electro-neutrality. In soils with base cations such as limestone, these ion pairs are calcium and magnesium, but in the areas where the rocks are resistant and the soils are lacking in base cations, the ion pair is aluminium which is present at a high level in rocks such as granite. The aluminium dissolves in the acid drainage and, if it exceeds a particular concentration, it kills fish by damaging the gills and the osmoregulatory mechanism which maintains the correct level of salts in the fish's bodily fluids. The aluminium also affects the larval stages of fish by reducing the action of an enzyme that dissolves the inner lining of the egg wall at hatching time, with the result that some larvae cannot emerge from their egg sacs. The result is that the number of fish fry is reduced and the population of fish slowly decreases. A characteristic of lakes affected by acidity is that the fish population is dominated by a reduced number of large fish, i.e. those that survived the hostile conditions, compared with a healthy lake where there are good representatives of all ages of fish.

It's not only the fish that are adversely affected by the low pH and high concentrations of aluminium. Earlier in this chapter, the impact of acidity on the microscopic diatoms in lakes was described. Another type of aquatic organism that varies in numbers according to the quality of the water, particularly in the flowing waters of rivers, is the invertebrates. They are discussed in more detail in Chapter 11 in relation to the impact of organic pollution on their numbers, but they can also serve as useful indicators of the acidity in rivers and streams.[5] One group of invertebrates that are especially sensitive to increases in acidity are those that make a shell, such as the snails and mussels, or an outer protective shell called an exoskeleton. Examples of the latter type are shrimps and water fleas. The

Table 19. Occurrence of selected invertebrates in acid streams

Mean pH	Group of invertebrates	Species absent
< 7.0	Crustacea	*Gammarus pulex*
< 6.0	Snails	*Lymnaea peregra*
		Ancylus fluviatilis
	Mayflies	*Baetis muticus*
		Caenis rivulorum
	Stoneflies	*Perla bipunctata*
		Dinocras cephalotes
	Beetles	*Esolus parallelepipedus*
	Caddis flies	*Glossoma* spp.
		Philopotamus montanus
		Hydropsyche instabilis
		Sericostoma personatum
< 5.5	Mayflies	*Baetis rhodani*
		Rhithrogena sp.
		Ecdyonurus spp.
		Heptagenia lateralis
	Stoneflies	*Perlodes microcephala*
		Chloroperla tripunctata
	Caddis flies	*Hydropsyche pellucidula*

reason why these invertebrates cannot survive in acid waters is because the low pH is usually associated with a lack of calcium and this element is an essential component for making a shell or exoskeleton. Shrimps and snails are rarely found in rivers where the pH falls below 5.5. There are other groups of invertebrates that also are affected by acidity to varying extents, such as the different types of mayfly larvae. Some of these can tolerate pH values as low as 4.5 whilst others cannot survive acidity less than pH 5.0 units. Table 19 shows the occurrence of some groups of invertebrates at different pH values.

The technique of collecting invertebrate species is described in Chapter 12, and it is thus possible to assess whether a particular upland river is affected by acidity, not only by the measurement of the pH, but also by examining the invertebrate animals present and seeing whether the absence of a sensitive group indicates that there have been 'pulses' of acidity brought in by acid rain clouds.

The cause and extent of acidification of rivers, lakes and lochs in the UK and elsewhere are now more clearly understood and some action has been taken to reduce the emissions into the atmosphere. At two of Britain's largest power stations the acid gases are neutralized before emission by passing them through fine limestone – so-called 'flue gas desulphurization' (FGD). Any future large power station must include FGD in its design. The amount of sulphur dioxide entering the atmosphere is slowly being reduced, as can be seen in Figure 27.

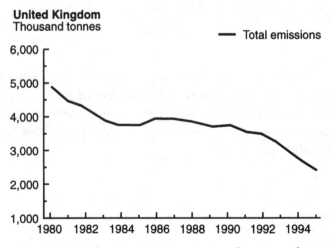

Figure 27. Reduction in SO_2 entering the atmosphere

Source: The Digest of Environmental Statistics No 19. *1997. The Environment in Your Pocket. Crown Copyright is reproduced with the permission of the Controller of Her Majesty's Stationery Office*

This reduction is partly because of the neutralization of the gas at the power stations but also because of the switch to fuels with less sulphur in them. Figure 28 shows the relative proportions of fuel used for producing power in the UK each year since 1970 and the reduction in the use of coal and the preference for natural gas and nuclear fuel is clearly seen.

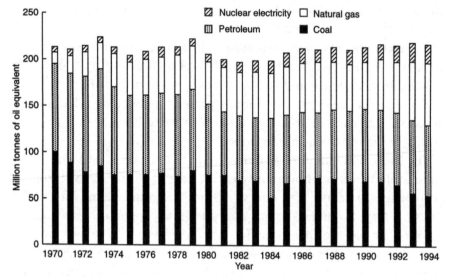

Figure 28. Proportion of fuels used for power generation in the UK, 1970–94.

Source: The Digest of Environmental Statistics No 19. *1997. The Environment in Your Pocket. Crown Copyright is reproduced with the permission of the Controller of Her Majesty's Stationery Office*

In contrast to the improvements in the control of acidity from power stations and other industrial sources, the amount of oxides of nitrogen is not declining and may even be slowly increasing. This is because of the steady increase in the number of vehicles on the roads. This is an issue dealt with in the next chapter

Notes

1. P. Elvingson and C. Agren, *Still with us. Although the pace may have slackened, acidification is still going on*, Swedish NGO Secretariat on Acid Rain, PO Box 7005, Göteborg, Sweden, 1996, (free publication).

2. R.W. Batterbee, A.C. Stevenson, B. Rippey, C. Fletcher, J. Natanski, M. Wik and R.J. Flower, 'Causes of lake acidification in Galloway, South West Scotland: A palaeo-ecological evaluation of the relative roles of atmospheric containment and catchment change for two acidified sites with non-afforested catchments.' *Journal of Ecology* 77, 1989, 651–72.

3. F.T. Last and R. Watling (eds), *Acid Deposition: its nature and impacts*, The Royal Society of Edinburgh, Edinburgh, 1991.

4. P.S. Maitland, A.A. Lyle and R.N.B. Campbell, *Acidification and Fish in Scottish Lochs*, Institute of Terrestrial Ecology, Grange-over-Sands, 1987.

5. B.R.S. Morrison, 'Acidification', in P.S. Maitland, P.J. Boon and D.S. McLusky (eds) *The Fresh Waters of Scotland*, Wiley, London, 1994.

Further reading

Acid News. A magazine with reports on acidification and air pollution. Available free, from Swedish NGO Secretariat on Acid Rain.

R.W. Edwards, R.S. Gee and J.H. Stoner, *Acid Waters in Wales*, Kluwer, Dordrecht, Netherlands, 1990.

D. Elsom, *Atmospheric Pollution: A Global Problem*, Basil Blackwell, London, 1992.

J. McCormack, *Acid Earth: The Global Threat of Acid Pollution*, Earthscan Publications, London, 1989.

9. Air pollution

Problems from air pollution have existed ever since the human race started to use fire. Anyone who has lit a bonfire or who has a coal fire at home will be well aware of the amount of smoke that is generated when they are first alight. Multiply the individual fires by the number of homes using them and it's easy to see why our towns and cities were such unhealthy places in the past.

The first documented complaints about air pollution can be traced back to 1257 when the wife of Henry III, Queen Eleanor, refused to stay in Nottingham Castle because of the choking air sent up to the royal chambers from the coal fires in the surrounding houses below the castle.

Hundreds of years passed before Parliament decided to act to improve air quality. Before then, the smoke and muck emanating from factories were signs of industrial progress. 'Where there's muck there's money' was an expression of the time. However, although a committee was appointed by Parliament in 1819 to investigate whether engines and furnaces could be operated without causing harm to health and comfort, no action was taken on its findings. Much later, the Public Health Act 1936 allowed local authorities to carry out inspections to detect emissions which were regarded as harmful to health, but again this measure was ineffective because the imminence of war required the maximum production of goods.

The milestone in air pollution prevention in the UK arose from the terrible smog which lasted for five days in London in December 1952. It occurred because the weather was particularly cold and still over the city which is situated in a river valley. A 'temperature inversion' took place which resulted in the warm air and fumes over the city being trapped beneath cold and denser air above it. The combination of dust, smoke and moisture resulted in a thick smog (the name was derived from a combination of 'smoke' and 'fog') which blanketed the area. Conditions were so bad that ducks from the parks crash-landed in the streets, buses were preceded along the streets by men with torches, and performances in the Sadlers Wells theatre were cancelled because the audience couldn't see the stage. These problems were trivial though when compared with the impact of the pollution on human health. Thousands of people suffered from breathing difficulties and the hospitals were full of casualties from the bad air. By the end of the event, it was estimated that 4,000 people had died as a result of the smog.

This episode spurred government action and a committee was appointed to investigate the causes and recommend action to prevent its repetition. The committee found that the major pollutants were particulates and gases, and proposed that new laws should be established to reduce the emissions of smoke grit and dust. An important part of the legislation was that it was be applied to domestic fires. The Clean Air Act of 1956 introduced controls over the height of chimneys and a new concept of 'smoke control areas'. In these areas, set up in towns, households were restricted to burning only fuel that did not emit smoke when lit, and people were given grants to help them convert their fireplaces to burn the 'smokeless' fuel.

The impact of this legislation was profound. For example, in Sheffield, an industrial city, the annual average amount of smoke and sulphur dioxide emitted today is only about one-tenth of the level in the early 1960s. Other changes have taken place over the past 40 years. There has been a large shift from coal to natural gas as a domestic fuel, and most houses have central heating or are warmed by electricity. Most of the electricity is now generated in large power stations situated well outside built-up areas and these are fitted with high stacks to allow maximum dispersion of the emissions. The modern power stations burn fuel more efficiently with the result that the amount of SO_2 emitted into the atmosphere has decreased. In 1979, an estimated 6.3 million tonnes entered the air from the UK's emissions but by 1990 this had fallen to 3.7 million tonnes. This reduction has led to an improvement in the quality of rainwater so that it is not as acidic as it used to be (as described in Chapter 8).[1]

Despite these overall improvements in smoke emissions, there is still public concern about urban air quality because of the increase in road traffic. Figure 29 shows the increase in the number of vehicles on Britain's roads, with the greatest use being on roads in built-up areas. For example, there are more households in the Greater London area that have two or three cars than anywhere else in the UK, but this is the most built-up

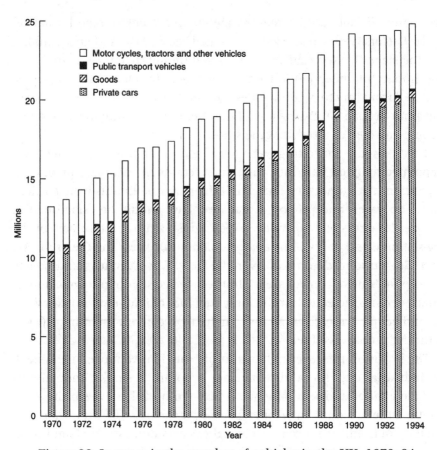

Figure 29. Increase in the number of vehicles in the UK, 1970–94
Source: The Digest of Environmental Statistics No 19. *1997. The Environment in Your Pocket. Crown Copyright is reproduced with the permission of the Controller of Her Majesty's Stationery Office*

area in the country. So these cars spend most of their time in the city streets, often proceeding at little more than running speed! By 1990 there were over 25 million vehicles on Britain's roads – a 28 per cent increase from 1980; 20 million of these vehicles are private cars, and their numbers are still growing rapidly (Cover Illustration 11).

The emissions from vehicles contain a complex mixture of pollutants and these vary according to the type of fuel. There has been a significant increase in the proportion of cars that have diesel engines. Although these are 'cleaner' because they have a better fuel consumption (more kilometres to the litre) and don't use petrol with lead in it, they emit more particles into the air. You are more likely to see black smoke coming from a diesel vehicle than from a petrol engine one. Between 1980 and 1990, there was a 75 per cent increase in the amount of black smoke emitted from road traffic whilst other pollutants, such as carbon monoxide, oxides of nitrogen and carbon dioxide, also increased substantially.

The main air pollutants in urban areas are: carbon monoxide (CO), oxides of nitrogen (NOx), hydrocarbons (HCs), sulphur dioxide (SO_2) and particulate matter – especially small particles which are sometimes called PM_{10}s because they are smaller than 10 μm. On sunny days, especially in summer, these gases interact with one another, particularly the nitrogen oxides and the hydrocarbons, to form ozone. This gas is harmful to human health and damages plants. High levels of ozone can cause breathing difficulties and are alleged to set off asthmatic attacks in vulnerable people. In the past ten years there has been a fourfold increase in the number of people suffering from asthmatic attacks in our cities. The Department of Health also estimates that pollutants in the air account for up to 20,000 hospital admissions and several thousand premature deaths each year.

Ozone is a very chemically active gas and reacts with other pollutants, such as the hydrocarbons, to form many so-called secondary air pollutants. One group of these secondary pollutants are known as the peroxy acetyl nitrates (PANs) which are particularly harmful to plants and humans.

The chemical reactions that take place in urban air on sunny days are complex and are still being studied. Some of the chemicals formed in the ultra-violet light from the sun are called free radicals and they are very reactive. PAN is formed by the reaction of the hydroxide radical on acetaldehyde which is present in vehicle exhausts:

$$OH + CH_3CHO \rightarrow H_2O + CH_3CO \text{ (acetyl radical)}$$
$$CH_3CO + O_2 \rightarrow CH_3COO_2 \text{ (peroxy acetyl radical)}$$
$$CH_3COO_2 + NO_2 \rightarrow CH_3COO_2NO_2 \text{ (peroxy acetyl nitrate (PAN))}$$

In August 1996, the UK Government decided to take action against air pollution, particularly that caused by motor traffic. It established targets for air quality that have to be achieved by the year 2005. These are shown in Table 20.

The quality of the air can be assessed in many ways. Sometimes, there are warnings about poor air quality on the weather forecasts on radio and TV. These warnings usually relate to the formation of ozone in built-up areas on calm sunny days. In Chapter 12, there are descriptions of simple methods of measuring air quality using filters or by measuring the yeasts

Table 20. Air quality targets for 2005

Pollutant	Target concentration
Benzene	5 parts per million
1,3-Butadiene	1 part per million
Carbon monoxide	10 parts per million
Lead	0.5 micrograms per cubic metre
Nitrogen dioxide	104.6 parts per billion
Ozone	50 parts per billion
PM_{10} particulates	50 micrograms per cubic metre
Sulphur dioxide	100 parts per billion

that live on leaf surfaces. These yeasts are sensitive to air quality and are killed off by pollutants. By carrying out a survey of the yeasts living on the leaves of trees in your local neighbourhood you should be able to find the most polluted areas and track down where the pollution is coming from.

There have been differing trends in air quality in our towns, cities and the countryside: some pollutants have increased whilst others have declined. For example, the amount of black smoke emitted from houses in urban areas is much less than 40 years ago because fewer people heat their homes with coal fires, and, for those who do, there are restrictions on the type of coal that can be burned. However, as mentioned earlier, the amount of black smoke from vehicles has increased greatly.

The amount of lead in the atmosphere has declined in recent years because many more vehicles use lead-free petrol and diesel for fuel, mainly because it's cheaper. This is a very good example of how the pricing of goods can influence people's buying habits with a benefit to the environment.

It was realized a few years ago that the lead in the atmosphere was increasing as a result of the rising numbers of vehicles on the roads. At that time, in order to make engines burn petrol more efficiently, a substance called tetraethyl lead (TEL) was added to petrol. TEL was an 'anti-knock' agent; in other words, it stopped the petrol igniting too early in the compression stage of the engine cycle. The lead from all the emissions from vehicles stayed in the atmosphere for a long time and spread slowly around the globe. It even contaminated the snow falling on Greenland: scientists who collected samples of snow from different layers in the glaciers there were able to show the increase in the amount of lead as a result of human activities.[2] Figure 30 shows the concentration of lead

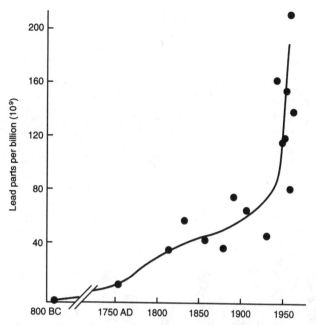

Figure 30. Increase in the concentration of lead in Greenland snow,
800 BC – 1965
Reprinted from Geochim. Cosmochim. Acta 33, *1969 with permission
from Elsevier Science*

at different depths (i.e. in earlier and later layers). Once the ages of the layers have been determined, the increases in lead concentration can be attributed firstly to the Industrial Revolution, when lead was smelted for a variety of uses such as lead pipes, and later to its use in the internal combustion engine.

This investigation was carried out in the late 1960s. Since then there has been a marked change in our use of leaded petrol. In the USA, it was banned, whilst in the UK, the price of leaded petrol was made more expensive than unleaded to persuade motorists to switch to the more 'environmentally friendly' fuel. Similar measures have been taken in other countries so that most cars now are using unleaded petrol. In the UK, the number of vehicles using unleaded petrol has increased greatly since it was first introduced in 1987, and this has resulted in a reduction in the amount of lead emitted (see Table 21). The effect of these changes has been remarkable. The air in our cities is less polluted by lead compounds, as shown in Figure 31 for two Scottish towns.

It is not only the local air that has benefited from these measures.

Table 21. Lead emitted by petrol engines in the United Kingdom, 1980–95

Year	Lead emitted (thousands tonnes)
1980	7.5
1985	6.5
1990	2.2
1993	1.5
1994	1.3
1995	1.1

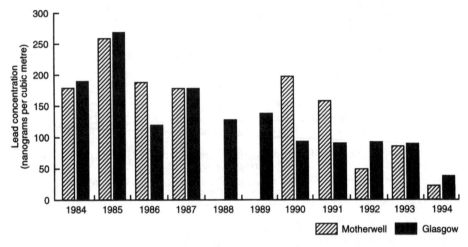

Figure 31. Concentration of lead in air in Motherwell
and Glasgow, 1984–94
Source: State of the Environment Report 1996, Scottish Environment
Protection Agency, Stirling, 1996

Further investigations have been carried out on the concentration of lead in the snow that has accumulated in Greenland since the 1960s.[3] As Figure 32 shows, the levels have fallen steadily and are now similar to what they were at the turn of the twentieth century!

This is a very good example of how improvements to the environment can be brought about. There was a great deal of concern about the air being polluted by lead from vehicle exhausts: it was even claimed that the intelligence of children living adjacent to busy roads was being adversely affected by the metal. Governments took action, either by banning the use of TEL in petrol or by making leaded petrol more expensive. Although these measures have reduced the lead levels in urban air, there is still concern over all the other pollutants that are emitted by

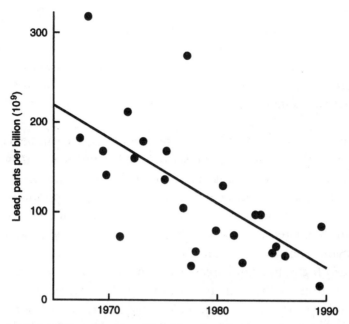

Figure 32. Concentration of lead in snow in Greenland, 1965–90
Reprinted by permission from Nature 353, *1991. Copyright Macmillan Magazines Ltd*

vehicles. Now, the Labour Government is hoping to improve air quality in our cities by trying to persuade us not to use cars so frequently, especially for going into towns and cities for work and shopping. The hope is that if car parking spaces are reduced and made more expensive, and the price of petrol is increased, people will find that it is cheaper and more convenient to use public transport rather than private vehicles.

Some cities in Europe and the USA have already achieved this. Trams have reappeared in Manchester, Sheffield and Paris, whilst trolley buses have been introduced in Los Angeles, Buffalo and Portland. Many cities in Holland and Germany retained their tramway systems but have recently modernized them. In Athens, a 2.5 km² area of the city centre has been made a car-free area: streets have been made into pedestrian areas, traffic re-routed and 'no-fare' buses started.

In California, a more radical method of reducing emissions is being tried out. The state has passed a law on zero emissions vehicles which applies to manufacturers that sell more than 35,000 vehicles in the state each year. By 1998, 2 per cent of the cars sold were supposed to be zero emission vehicles, and this is set to rise to 5 per cent by 2001 and 10 per cent by 2003. This has spurred car manufacturers to produce more battery-powered vehicles and to improve the technology so that they are speedy and have sufficient battery storage to last a whole day.

The concern about air quality relates to a number of chemicals and these were listed on page 90. One of these, ozone, is a 'Jekyll and Hyde' gas. If present in the air we breathe it is a pollutant, but if it is present in the upper atmosphere it protects us from harmful ultra-violet (UV) radiation.

In the lower atmosphere – the troposphere – ozone can be damaging to human health, farm crops, forests and wild plants. It is usually formed on calm summer days because of the presence of sunlight and polluting gases, such as oxides of nitrogen and sulphur. The concentration of tropospheric ozone in industrial areas doubled in the period 1940 to 1970 in line with the increase in the number of vehicles but, because of chemical reactions which take place in the atmosphere, concentrations of ozone are usually highest in rural areas. This is because in towns and cities the pollutants from vehicles, such as various hydrocarbons, react with the ozone and reduce its concentration, but in adjacent country air these are absent and so ozone levels increase. From the monitoring results of air quality in different parts of England for 1997, it can be found that the health standards for ozone were exceeded on over 60 days in Lullington Heath in Sussex, on 37 days in Teddington, London, and 27 days at High Muffles in North Yorkshire. On a typical early summer day in 1998, the rural ozone concentration in south Scotland was 45 parts per billion whereas the concentration in Edinburgh was 19 parts per billion.

The combination of all the different urban pollutants can produce the problem of so-called 'photochemical smogs'. These are formed in cities because of the high numbers of vehicles emitting gases which interact to form a complex mixture. There have been a number of serious incidences of photochemical smogs in cities in the USA, particularly Los Angeles because of its sunshine and large numbers of vehicles. At one time these types of smogs were considered to be a problem only in the USA but in recent years levels which are hazardous to human health were recorded in a number of cities in Europe. There was a serious episode in Athens in August 1984 when an acid smog resulted in more than 500 people being taken to hospital with breathing difficulties. More recently, the summers of 1995 and 1996 resulted in high concentrations of ground-level ozone. According to a report from the European Topic Centre on Air Quality,[4] the threshold level for protection of human health (110 µg per cubic metre as an eight-hour average) was exceeded in all EU member states in 1995 for 1–2 consecutive days. The longest episode lasted eight days. It also reported that 78 per cent of the EU urban population was exposed to levels of ozone above the threshold level for at least one day of that year and that 9 per cent suffered from excessive levels for 50 days.

Notes

1. *Digest of Environmental Statistics 1994*, Dept of Environment, London, 1994.

2. M. Murozumi, T.J. Chow and C.C. Patterson, 'Chemical concentration of pollutant Pb aerosols, terrestrial dust and sea salts in Greenland and Antarctic snow strata', *Geochim. Cosmochim. Acta*, 33, 1969, 1247.

3. U. Gorlach, J.P. Candelone and C. Boutron, 'Changes in heavy metals concentration in Greenland snow during the past twenty years', in *Heavy Metals in the Environment*, International Conference, Edinburgh, CEP Consultants, Edinburgh, 1991.

4. European Topic Centre on Catalogue of Data Sources founded by the European Environment Agency, Copenhagen, Denmark.

10. Global warming

In recent years, the temperature of the earth's atmosphere has been warmer than at any time since measurements were first taken in 1860. Overall, it is estimated that the average overall temperature increase is between 0.3 and 0.6°C. In April 1997, it was announced that satellite measurements of the northern hemisphere of the globe showed that spring was arriving seven days earlier than ten years before and that the leaf fall of autumn was taking place four days later. These were just two examples of global warming caused by the so-called greenhouse effect.

The way the greenhouse effect is reported in the media, it would seem that it's a bad thing but in fact it's what keeps our planet habitable. If it wasn't for the presence of carbon dioxide and other gases in the atmosphere above the earth, we would be much colder – probably about –18°C! The greenhouse effect is essential to life on earth.

The greenhouse effect is the mechanism whereby heat from the sun reaches the atmosphere, and the layer of gases allows the short-wave radiation to pass through and warm the air and the earth's surface – just like the glass in a greenhouse. Some of the long-wave radiation (the infrared) given off by the warming process returns to space but most is retained because it cannot pass through the gas layer. The process is summarized in Figure 33.

The main gases responsible for causing the greenhouse effect are carbon dioxide, water vapour, methane, nitrous oxide and chlorofluorocarbons (CFCs). The problems arise when the levels of these gases increase in the atmosphere and make the greenhouse effect stronger. It's like putting thicker glass into a greenhouse and making it more efficient at trapping heat.

Of particular concern is the increase in the amount of carbon dioxide. As a result of human activities, the growing population and our increasing use of fossil fuels, the amount of carbon dioxide entering the atmosphere is rising. The concentration of CO_2 is 8 per cent greater now than in 1960

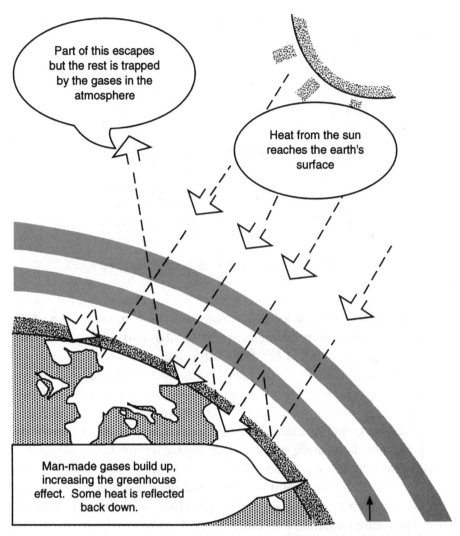

Greenhouse gases ("glass")

Figure 33. The effect of greenhouse gases on warming of the earth's surface by the sun's rays
Source: Global Warming; the Greenpeace Report, *by permission of Oxford University Press*

and 25 per cent more than in the early 1800s. The rate of increase is such that, if we carry on as we are, the concentration will double from present-day levels in the next 100 years.[1]

For thousands of years, there was a balance between the greenhouse gases that were produced naturally (with the exception of CFCs which are artificial substances) and their absorption by oxidation and photosynthesis. Now it appears that the balance is upset and the gases are being produced faster than they can be utilized. Figure 34 shows the increase in input of carbon dioxide into the atmosphere in the past 100 years.

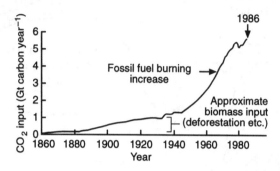

Figure 34. Increase in carbon dioxide discharged into the atmosphere, 1860–1986

Although there is a great deal of concern about the warming of the planet, the temperature has been at least as high in the past as it is now. There are various ways of obtaining information on the temperature of the earth and the concentration of carbon dioxide in the atmosphere from thousands of years in the past. For example, Swiss scientists have collected air bubbles trapped in polar ice caps that have been there for centuries and they have analysed their carbon dioxide content.[2] The layers of the ice have been dated using radioactive dating techniques based on the ratio of the different oxygen isotopes. This information is then correlated with estimates of the earth's temperature at the time which can be obtained from fossil records or looking at different types of pollen. The results of this type of investigation are shown in Figure 35.

If there is a true correlation between the temperature of the earth and the amount of CO_2, then a doubling of the amount of greenhouse gases will give rise to an overall temperature increase of between 1.4 and 4.5°C.

There is still some controversy between scientists about global warming. Although an international panel of experts have assessed all the available data and accepted that global warming is happening, there are others who say that the increase in temperature is nothing more than the natural recovery from the cool of the so-called Little Ice Age which occurred between 1550 and about 1850. It has also been said that the earth will respond to the increase in CO_2 because it will promote the photosynthesis of plants. After all, returning to the observation about

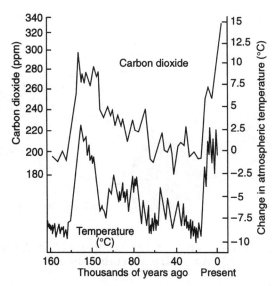

Figure 35. Changes of temperature and CO_2 over time

earlier springs and later autumns made at the beginning of this chapter, the early arrival of greenery on trees and plants and its prolonged presence means that more of the atmospheric CO_2 is being used up by plants. A recent book has put forward an alternative to the 'greenhouse gas theory' for global warming.[3] It is based on the interplay between cosmic rays and the fluctuating output from the sun. Cosmic rays help in the formation of the earth's cloud cover which causes a cooling of the earth, whilst when the sun is active solar winds reduce the cosmic rays. The decrease in the cosmic rays reduces cloud cover and the earth warms up. It is proposed that the variations of these two factors explain not only the climate change observed now but also that of past aeons.

Another factor to consider is the movement of the earth around the sun. In the last million years, there have been regular changes in the earth's temperature, with ice ages lasting for about 100,000 years followed by warmer spells lasting about 10–20,000 years. Because of the change in the angle that the sun impinges on the earth there have been ten such cycles in just over one million years and, at the present time, we are nearing the end of a warm spell that began 10,000 years ago. These patterns of temperature change have been studied by astronomers and have been found to be linked to changes in the earth's movement round the sun[4] as shown in Figure 36. Although these natural changes of temperature are known about, the scientists studying global warming say that

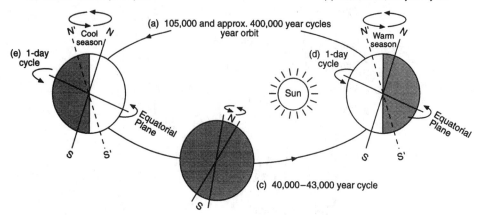

Figure 36. The tilts and wobbles in the earth's circuit round
the sun which affect the climate

the rate of temperature increase is faster than would have occurred in
the normal course of events.

One of the most important factors to take into account in the relationship between temperature change and the increase in greenhouse
gases is the role of the oceans. These cover 70 per cent of the earth's
surface and this salt water acts like a giant thermostat, by absorbing heat
from the atmosphere when the air is warm and then releasing it when
the air gets cold. It also absorbs carbon dioxide as this first dissolves in the
water and is then utilized by phytoplankton in their photosynthesis. This
exchange between the air and the oceans takes place on a huge scale –
involving far greater amounts than those emitted to the atmosphere from
human activities, as shown by the following figures:

Carbon dioxide absorbed by the oceans each year = 105 billion tonnes
Carbon dioxide released by the oceans each year = 102 billion tonnes
Carbon dioxide emitted into the atmosphere from human activities =
    6 billion tonnes

From these figures it is clear that the fate of any excess carbon dioxide in
the atmosphere is influenced by the number of phytoplankton in the
surface layers of the oceans. For example, it has been estimated that the
amount of phytoplankton in the oceans at the time of the last ice age was
nearly 50 per cent more than today.[5] The photosynthetic activity of these

organisms reduced the amount of carbon dioxide in the atmosphere and this in turn allowed more infra-red radiation to escape from the earth (the greenhouse 'glass' was thinner). The result of all these interactions was that the temperature of the earth was about 10°C cooler than it is now.

The fate of the increased CO_2 being released into the atmosphere as a result of human activities depends a great deal on the number of phytoplankton in the world's oceans. Not much is known about any changes in their numbers but there is a lot of research taking place. The problem is the lack of information from thousands of years ago when obviously no one was taking systematic measurements. It could be that the increased amount of CO_2 will activate the growth rate of the phytoplankton and this will then result in a reduction in the atmospheric CO_2. There are other factors to consider, one being the absorption of heat by the oceans. As the earth's temperature increases, the ocean waters also slowly warm up as a result of absorbing this heat and the circulation patterns distribute the heat to different parts of the globe. Another factor is the influence of cloud cover. The increased temperature may give rise to more clouds because of the higher rate of evaporation of sea water. Clouds reflect sunlight and therefore less solar radiation reaches the ground, with the result that not as much heat is absorbed by the earth. On the other hand, water vapour is a 'greenhouse gas' and helps retain heat, although it is estimated that clouds are twice as effective at reflecting sunlight back into the atmosphere as they are at retaining heat.

You will realize from this that there are many arguments about global warming and we still have much to learn. All we have at the moment are predictions and some evidence that the earth is getting warmer, and we have to act on this information.

The best estimates that we have are that the amount of CO_2 in the atmosphere will double in the next hundred years and that this will result in an increase in temperature of 4.2°C. What is the likely effect of this in northern Europe? Table 22 sets out some likely scenarios.

These predictions, if correct, will make a profound difference to life in northern Europe. An indication of how great a change may occur can be obtained by looking at historical records of the early Middle Ages (AD 800–1200) when it seems that the average temperature was 1°C higher than at present. At that time, it was possible to grow corn as far north as Iceland, there were trees growing in northern Scandinavia (now it is only mossy tundra) and in southern England, apricots, peaches and grapes were grown outdoors.

Table 22. Climate change in northern Europe predicted from a doubling of atmospheric CO_2

Climate parameter	Climate change
Mean annual temperature	Increased by 4–5°C
Mean winter temperature	Increased by 5–6°C
Mean summer temperature	Increased by 2–3.5°C
Length of growing season	Increased by 70–150 days
Length of winter (temperature below freezing)	Decreased by 2–4 months
Length of summer (temperature above 10°C)	Increased by 2–3 months
Mean annual rainfall	Increased by 10–50 per cent

Obviously with temperature rise of over 4°C, the effects on agriculture will be far more profound and southern Britain would have a climate like the Mediterranean countries today. There would also be much more productive land in northern Scandinavia and Russia. This might seem very desirable, but there is also a marked disadvantage to this predicted global warming. Experts expect that there would be an increase in the rate of desert formation in large areas of the USA, China, North Africa and other countries where agriculture is already difficult because of infrequent rainfall.

Another prediction about global warming is of a general change in our climate. It is expected that, overall, rainfall will increase but summers will be drier. This means that winters will be much wetter and the rain will be accompanied by more ferocious storms which will cause structural damage as well as localized flooding. Some evidence of the change in rainfall in Scotland is obtained by looking at the flow rates of rivers. Figure 37 shows the change in the average flow rate of the River Clyde over the 30-year period, 1966–96.

This steady increase in average flow rate conceals another effect and that is that the low flows of summer have not changed. In other words, the increased average flows have occurred because of more rainfall in winter. Heavy rain in the winters of 1994 and 1996 gave rise to severe and damaging floods in the West of Scotland with some of the highest ever flow rates in the rivers. Since these flow data were published, more measurements have been made of the flow of the Clyde. The upward trend in average flow rate has declined, but this can be attributed to some particularly dry summers, especially those of 1995 and 1996.

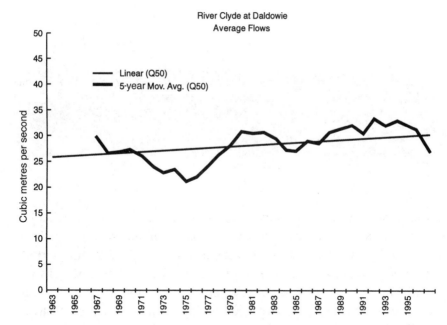

Figure 37. Change in the average flow rate of the River Clyde, 1963–95

Notes

1. A. Nilsson, *Greenhouse Earth*, Wiley, Chichester, 1992.
2. H. Oeschger and I.M. Mintzer, 'Lessons from the ice cores: rapid climate change during the last 160,000 years', in M. Mintzer (ed) *Confronting Climate Change: Risks, Implications and Responses*, Cambridge University Press, Cambridge, 1992.
3. Nigel Calder, *The Manic Sun – Weather Theories Confounded*, Pilkington, London, 1997.
4. W.R. Peltier, 'Our fragile inheritance', in C. Mungall and D.J. McLaren (eds) *Planet Under Stress: The Challenge of Global Change*, Oxford University Press, Oxford, 1990.
5. *Oceans and the Global Carbon Cycle*, Biogeochemical Ocean Flux Study (BOFS), BOFS Office, Plymouth Marine Laboratory, Plymouth, 1989.

Further reading

K.T. Pickering and L.A. Owen, *An Introduction to Global Environmental Issues*, Routledge, London, 1994.

D.M. Gates, *Climate Change and its Biological Consequences*, Sinauer Associates Inc., Sunderland, Mass., 1993.

11. Biological indicators of the quality of the environment

In this book, various types of pollution have been described which affect the quality of the air, land and water. The pollutants can usually be measured by analytical chemistry and, in Chapter 12, different methods of chemical analysis are described. However, chemical analysis tells you only the amount of the pollutant present and nothing about its effect on the environment. For this, we need biological methods because living organisms are exposed to pollutants and react according to the length of the exposure and their sensitivity to the pollutant. However, before we can assess whether the organisms are affected by a pollutant, we need first of all to look at an unaffected community. This is an essential part of any scientific investigation, whether it be testing a new medicine or a weedkiller: you have to know what is the state of the unaffected environment before you can draw any conclusions about the impact of the pollutant.

When studying pollution of the environment, the usual technique is to look at those species most sensitive to pollution. For example, for assessing air pollution, there are numerous plants and trees that can tolerate a variety of pollutants at quite high levels, as is shown by the numbers of trees and flowers in city centres where the air is of poor quality because of traffic fumes. However, the lichens that grow on stonework are much more sensitive and there are few, if any, in polluted city centres, although they are numerous in clean air areas. The lichens are known as indicator organisms because a study of the number and variety present gives an indication of the quality of the air.

When assessing water pollution, there are various aquatic organisms that can be studied – fish, water weeds, plankton and invertebrates. The variety of fish life, however, is limited and they can often swim away from pollution when it occurs. The water weeds cannot move away to avoid pollution but they usually die off in winter so can be studied only during summer months. The study of plankton is a very specialist subject so they are not considered to be suitable indicator organisms. The invertebrate life is very diverse, they have limited movement and so are exposed to changes in the water quality. Additionally different species have different tolerances to pollutants. In all, they make ideal indicator organisms. They are easy to catch and be identified, and there is a great deal of information available about the effects of various pollutants on them.

The major part of this book has been describing various aspects of water pollution. This is because it is the sector of the environment that is most affected by pollutants. Even when we considered the effects of emissions of acidifying gases into the atmosphere, it was the rivers and lakes which showed the adverse effects. Likewise in the section on solid waste, the main concern is the pollution of water by substances being washed from the waste into drainage water. The agencies responsible for protecting the environment from pollution spend most of their time and effort on water issues and we also will be looking largely at methods of investigating water pollution.

For a biological assessment of the effect of pollution on aquatic life we must first look at an unpolluted environment. All life on earth depends ultimately on the energy from the sun. This energy is absorbed by plants and used, by the process of photosynthesis, to convert carbon dioxide and water into carbohydrates as shown by the familiar equation:

$$6CO_2 + 6H_2O \xrightarrow{\text{sunlight}} C_6H_{12}O_6 + 6O_2$$

This process is known as primary productivity. The organisms that do this are known as autotrophs: they are the primary producers of food in the world and the great majority of all other organisms depend upon them. They convert solar energy into chemical energy. This chemical energy is released when the organism that feeds on the autotrophs breaks down the carbohydrates and uses the energy stored in it for its life processes.

In the aquatic environment, the autotrophs are the water plants, algae and some bacteria but an important part of the primary productivity comes from the vegetation at the edge of rivers such as the trees, bushes and tall plants. The plants that grow at the edge of water are known as riparian plants and the 'bits' that fall off them into the water (particularly in autumn at leaf fall) adds to the productivity of the water. This addition of organic material to the water is called detritus and is an important food source for a variety of aquatic life on which other organisms prey.

The detritus is broken down and consumed by bacteria, fungi and various invertebrates (known as detritivores). The bacteria are essential for the health of the river community because they provide food for other organisms. The bacteria are consumed by protozoa and these in turn are eaten by various invertebrates such as flat worms, crustacea and molluscs (snails). The crustacea and worms are consumed by insect larvae and fish, whilst the fish also eat water weeds and insect larvae. Finally, the fish are preyed on by birds such as herons and kingfishers, as well as by anglers. This whole system of interconnected producers and consumers is known as the food web and, in an unpolluted environment, there is a balance between all the components that make up the web, as shown in Figure 38.

One important factor which determines the types of organisms present in a stream is the habitat, in other words the place where they live. The water, mud, weeds and the surfaces and undersides of stones are all quite

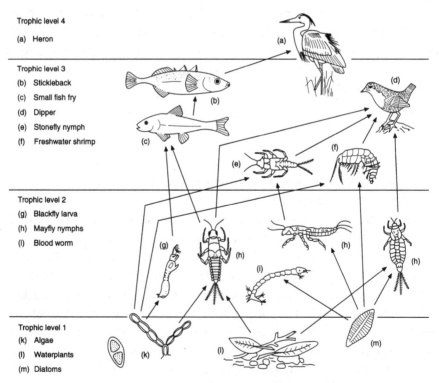

Figure 38. An example of an aquatic food web

different habitats which are occupied by various species. If we consider a stone sitting on the river bed, there are variations in the speed of currents of water flowing over it as shown in Figure 39.

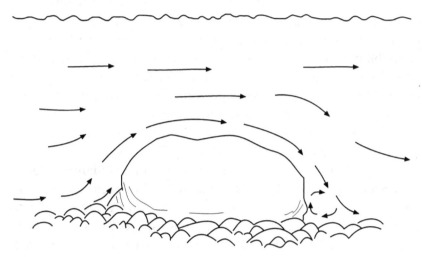

Figure 39. Variations in the speed of currents over a stone in a stream

Those organisms that live on the surface of the stone have to be able to cling on in the strong current. The mayfly larvae do this by having flattened streamlined bodies and sharp claws, whilst the buffalo gnat larvae spin a web over the stone surface and 'glue' themselves to it with a basal pad of many claws. The undersurface of the stone has quieter water and is occupied by water hog lice and shrimps which crawl around the stream bed and eat detritus. Another habitat is the muddy bed of the river in the slower moving stretches. Here, the invertebrates most suited to this environment are those that burrow into the mud, such as freshwater mussels and midge larvae.

In the balanced community, if one component of the food web increases then it is brought under control by an increase in the number of organisms that feed on that component. Eventually the number of consumers exceeds the food supply, so they die off either because they run out of food or because there is an increase in the number of their predators. For example, in autumn there is an increase in the supply of detritus, so the numbers of bacteria increase which in turn improves the productivity overall in the community. The food supply is reduced when the leaves have been broken down and this then results in an overall decline in the numbers of organisms.

The balance of the community can be markedly upset by pollution. Pollution was defined in Chapter 1 and included in that definition was 'The discharge... of substances or energy... the results of which are such as to cause... harm to aquatic life...' The discharged substances can be harmful by being toxic or causing an imbalance in the aquatic community. Poisonous discharges will kill the organisms according to their susceptibility to the poison or their ability to escape from the polluted water. There are many toxic substances which can enter streams, such as pesticides, toxic metals like copper and lead, and acids. Most often, though, pollution of water is caused by excessive organic matter. This can originate from farm waste, inadequately treated sewage or industrial waste from, say, a brewery or food factory.

The large input of organic matter provides an abundance of food for bacteria and they multiply at a great rate. The food web is upset and the variety of organisms react to this upset. The large number of bacteria break down the waste and, at the same time, utilize the oxygen dissolved in the water. This depletion of the dissolved oxygen is counteracted by more oxygen dissolving in the water from the atmosphere. This process is helped if the river is fast flowing or there are plenty of waterfalls. However, if the river flow is sluggish or the demand for oxygen by the bacteria exceeds the input from the atmosphere, the dissolved oxygen will be in short supply. Those organisms sensitive to depleted oxygen levels will move away by either swimming (fish) or drifting (invertebrates), whilst some will be killed. Other organisms though have adapted to fluctuating oxygen levels. In particular, there are some invertebrates which have haemoglobin in their bodies and are able to store oxygen for use in times of shortage. Examples of these are the midge larvae (also known as blood worms because of their red colour) and tubificid worms.

So, in rivers that are badly polluted with organic waste, the invertebrate population is dominated by masses of worms and little else, whilst a clean river has a wide variety of organisms with no one species dominating. In between these two extremes there is a gradation of organisms that have different sensitivities to oxygen depletion. In a river which is only mildly polluted and which is fairly fast flowing, the input of extra organic matter increases the biomass (the total number of organisms). Those invertebrates that consume organic matter thrive in the favourable conditions because of the extra food supply, so in these waters there are plenty of shrimps and water hog lice as well as the worms.

Freshwater biologists use this variation to assess the quality of water.

They collect representative samples of the invertebrate community and examine the number and variety that are present. From this examination they can produce a Biotic Index, which is a number expressing the water quality: the higher the number the cleaner the water. One of the most regularly used Biotic Indexes in the UK was devised by a group of biologists and is known as the Biological Monitoring Working Party (BMWP) score. In this scheme, each invertebrate family is assigned a number depending on its sensitivity to pollution: a stonefly larva requires clean water with a high concentration of dissolved oxygen so has a score of ten, whereas a blood worm will have a score of one. The scores for all the various families present are added to produce the BMWP score. Table 23 gives examples of the invertebrate populations, with their Latin names,

Table 23. Invertebrate families and BMWP scores for four rivers of different levels of pollution

River	Stretch sampled	BMWP score	Family	Score
Kittoch Water	downstream East Kilbride	27	*Baetidae* (mayfly larvae)	4
			Simulidae (blackfly larvae)	5
			Chironomidae (midge larvae)	2
			Gammaridae (shrimp)	6
			Asellidae (water hog louse)	3
			Erpobdellidae (leech)	3
			Lymnaeidae (snail)	3
			Oligochaeta (worm)	1
River Kelvin	near source	50	*Baetidae*	4
			Simulidae	5
			Gammaridae	6
			Hydropsychidae (caddis fly larvae)	5
			Lymnaeidae	3
			Tipulidae (crane fly larvae)	5
			Limnephilidae (snail)	3
			Muscidae (house fly type larvae)	(no score)
			Planariidae (flat worm)	5
			Hydrobiidae (snail)	3
			Physiidae (snail)	3
			Sphaeriidae (bivalve snail)	3
			Oligochaeta	1
River Irvine	middle reaches	100	*Baetidae*	4
			Chironomidae	2
			Gammaridae	6
			Lymnaeidae	3

Table 23 (continued)

River	Stretch sampled	BMWP score	Species	Species score
River Irvine (continued)			Tipulidae	5
			Planariidae	5
			Asellidae	3
			Hydrobiidae	3
			Physidae	3
			Oligochaeta	1
			Erpobdellidae	3
			Leuctridae (stonefly larvae)	10
			Perlodidae (stonefly larvae)	10
			Ephemerellidae (mayfly larvae)	10
			Heptageniidae (mayfly larvae)	10
			Dytiscidae (beetle)	5
			Elmidae (beetle)	5
			Hydracarina (mite)	(no score)
			Glossiphoniidae (leech)	3
			Ancylidae (limpet)	6
			Sphaeriidae (mussel)	3
River Doon	middle reaches	132	Baetidae	4
			Chironomidae	2
			Gammaridae	6
			Asellidae	3
			Erpobdellidae	3
			Tipulidae	5
			Limnephilidae	7
			Simuliidae	5
			Leuctridae	10
			Oligochaeta	1
			Ephemerellidae	10
			Leptoceridae	10
			Heptageniidae (mayfly larvae)	10
			Caenidae (mayfly larvae)	7
			Ryacophilidae (caddis fly larvae)	7
			Glossosomatidae (cased caddis fly larvae)	7
			Hydropsychidae (caddis fly larvae)	5
			Brachycentridae (cased caddis fly larvae)	10
			Empididae (fly larvae)	(no score)
			Rhagionidae (snipe fly larvae)	(no score)
			Elmidae (beetle)	5
			Aphelocheiridae (saucer bug)	10
			Hydrobiidae (snail)	3
			Ancylidae	6
			Sphaeriidae	3

for four rivers in Scotland with different degrees of pollution.

The examination of invertebrate stream life need not be as complicated as this. A great deal of skill and experience are required to identify all the different invertebrate species in order to produce a biotic score. A rough estimate of the water quality can be obtained by collecting a representative sample of the stream bed community and seeing which groups are present and which are absent. You will notice in Table 23 that the higher the biotic score, the greater the variety of invertebrates present.

These results are all for rivers in Scotland where the upper limit for BMWP scores is close to 200. In the chalk rivers of southern England, there is an even greater diversity of invertebrate life and the maximum BMWP scores can be much higher.

Further reading

A. Cadogan and G.A. Best, *Environment and Ecology*, Nelson Blackie, Bishopbriggs, 1992.

J.M. Hellawell, *Biological indicators of Freshwater Pollution and Environmental Management*, Elsevier, London, 1986.

D.M. Roenburg and V.H. Resh (eds) *Freshwater Biomonitoring and Benthic Macroinvertebrates*, Chapman Hall, London, 1993.

R. Fitter and R. Manuel, *Collins Field Guide to Freshwater Life*, Collins, Glasgow, 1986.

12. Measuring the quality of the environment

This book has described a variety of environmental pollutants and their effects on people and other organisms. In laboratories throughout the world, professional scientists are regularly collecting samples from all sectors of the environment and analysing them for pollutants. Sometimes the analytical results are presented in court to prosecute polluters, whilst in other laboratories the data are used to find safe levels so that the environment is not adversely affected. In most laboratories, the analytical instruments used are complex and expensive, and are capable of rapidly measuring trace quantities of pollutants. It is possible, however, to obtain an indication of the quality of air, water and land with simple techniques and some of these are described in this chapter. Suggestions are made as to some investigations that can be carried out in school or college laboratories.

One important point to make is that such investigations should be kept small and simple to begin with. It can be very frustrating to set out with an idea of, say, studying a range of pollutants in various rivers and then to find that you run out of time and have few results to complete a project. Start off small and build on the project as experience is gained. For example, you could assess the impact of a discharge on a stream by collecting samples upstream and downstream of the effluent and of the effluent itself, and analysing the samples for three or four types of chemicals. If time permits, the project could be extended by looking at the variety of invertebrates upstream and downstream, or extending the number of chemical tests. You could collect samples during different flow rates and assess the effect of dilution on water quality.

Sewage effluent

In Chapter 2, the chemistry of sewage effluent was described and examples were given of the variety of different chemicals in good and poor effluents and in the receiving streams. There are relatively simple analytical measurements for the determination of some of these parameters, using equipment and chemicals usually available in school and college laboratories. This section describes the analytical procedures for the measurement of suspended solids, biochemical oxygen demand (BOD), dissolved oxygen, permanganate value (PV), pH, chloride and alkalinity. Each of these tests will yield valuable information on the quality of a river and the effect of a discharge on it.

Suspended solids

This is a simple but important measurement of the amount of silt in suspension in the water. The pollution control authorities often restrict the amount of suspended solids that can be present in a discharge. Suspended solids can be a particular problem resulting from quarries, building sites, new roads and ploughed land. Excessive amounts reduce the penetration of light into the water and this in turn can restrict the photosynthesis of aquatic plants. The suspended silt eventually sinks to the river bed and too much silt will clog the gaps between stones and will thus change the habitat for invertebrates and other water creatures.

Apparatus required:
    Oven regulated to about 105°C
    Water pump to provide suction
    1 litre filter flask
    Hartley funnel (see Figure 40)
    Measuring cylinder
    Glass-fibre filter paper (preferably grade C) of a size compatible with the filter funnel
    A pair of tweezers
    Desiccator
    An accurate balance capable of measuring to 0.1 mg

Procedure
A clean glass-fibre filter paper is labelled with a number written in pencil

and placed in the oven at 105°C for at least an hour. Next it is put into the desiccator to cool and then weighed accurately (weight A). The paper should be picked up with tweezers to avoid it being handled and picking up sweat from fingers which will change its weight. The dried filter paper is then put into the Hartley funnel on top of the perforated disc.

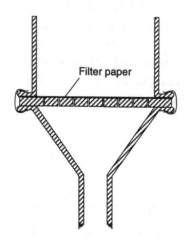

Figure 40. Hartley funnel and filter paper

A pre-determined volume of one of the collected samples is mixed well and measured into the measuring cylinder. The water pump (or vacuum pump) is started and the sample then poured carefully into the funnel. When all the water has passed through the filter paper, the filter pad is rinsed with a little distilled water, taken off the filter funnel and placed back in the oven for at least two hours. After cooling in the desiccator the filter paper is re-weighed: this is weight B.

$$\text{The amount of suspended solids, expressed as mg/l (ppm)} = \frac{(B - A) \times 10^6}{\text{volume of sample}}$$

The preferred volume of sample to take for this test depends on the amount of suspended solids present. If the water appears very clean, then a larger volume is required for filtering in order to obtain enough solids on the filter paper to make a weighable amount. By contrast, for a very turbid sample, only a small volume is required, otherwise it will take a long time to pass through the filter paper. As a rough guide, the following volumes are suggested for different types of samples:

Sample	Suggested volume (ml)
Clean river water	250–500
Dirty river water	100–250
Clean sewage effluent	100–250
Poor sewage effluent	50–100
Untreated sewage	50
Turbid, silty water	<50

Some laboratories will not have an analytical balance of the accuracy and sensitivity required (measurable to 0.1 mg). In this case, an indication of the suspended solids can be obtained by measuring the turbidity of the sample using a simple measuring technique.

Turbidity

The turbidity of the water is a measure of its clarity. Turbid water is difficult to see into because the suspended particles disperse light.

Apparatus required:
A piece of glass tubing about 1 m long and 2.5 cm in diameter. To this is cemented, using superglue, a piece of plain glass onto which is drawn a black cross of lines about 1 mm wide.

The sample is poured into the cylinder until it is just not possible to see the cross when viewed from above. The depth of water in the cylinder is measured.

A rough classification of the water quality is as follows:

Clean water	–	water depth is over 600 mm
Slightly polluted water	–	about 300mm
Polluted water	–	less than 100 mm.

Chloride

The amount of chloride in a water sample can give an indication of the amount of sewage effluent in river water. As shown in Tables 3 and 4 in Chapter 2, sewage effluent contains about 50–100 mg/l of chloride ion. This originates largely from human urine which typically has 1 per cent chloride ion present. Clean river water on the other hand usually has <10 mg/l chloride so the measurement of this ion can give an indication of the amount of dilution that an effluent receives in a river.

A special problem arises with chloride in winter months when roads and footpaths are treated with a mixture of salt and grit to free them of snow and ice. When a thaw arrives, the salt is washed into drains and thence into nearby streams giving rise to high values of chloride.

Apparatus required:
Porcelain basin of 200 ml capacity
Glass stirring rod
100 ml measuring cylinder
50 ml burette
1 ml dropper

Reagents:
Silver nitrate solution. Dissolve 4.791 g silver nitrate (previously dried at 105°C) in distilled water and make up to 1 litre.
Potassium chromate indicator. Dissolve 5 g potassium chromate in 100 ml of distilled water. Add silver nitrate solution to it until there is a slight red precipitate. Filter into a stoppered bottle.

Procedure
100 ml of sample is measured in the cylinder and then poured into the porcelain basin. Add about 1 ml of potassium chromate indicator using the dropper. Titrate the sample with the silver nitrate solution, constantly stirring the sample. As the silver nitrate goes into the sample, it forms a red colour but this changes to yellow on stirring. This is because the silver nitrate first reacts with the chloride ion to form insoluble silver chloride. Once all the chloride has been precipitated, it forms insoluble silver chromate which is red. The end point of the titration is reached when the colour of the precipitate just changes from yellow to an orange-red.

Chloride content of sample (mg/l) = [volume of silver nitrate] − 0.2 × 100
(The 0.2 is subtracted to allow for the precipitation of silver chromate.)

Alkalinity

The alkalinity of water is that which neutralizes 0.1M acid when titrated to an end point with methyl-orange indicator. The acid neutralizes the carbonates, bicarbonates and hydroxides in the sample, largely the first two components. As described in Chapter 2, these are the ultimate oxidation products of organic matter when it is purified by bacteria. The results of a survey carried out by collecting water samples along the length of a river which receives effluents at different places, would show is a steady increase in the alkalinity value along the river.

Apparatus required:
  Porcelain basin of 200 ml capacity
  Glass stirring rod
  100 ml measuring cylinder
  50 ml burette

Reagents:
  Hydrochloric acid (0.1M). Dilute 10 ml of concentrated hydrochloric acid to 1 litre with distilled water. This acid solution should be standardized with 0.05M sodium carbonate solution and then its volume adjusted so that it is exactly 0.1M. Alternatively, a pre-standardized solution can be purchased.
  Methyl-orange indicator. Dissolve 0.5 g methyl orange in 1 litre of distilled water.
  Sodium carbonate solution (0.05M). For standardizing acid. Dissolve 5.30 g of sodium carbonate (dried at 105°C) in distilled water and dilute to 1 litre.

Procedure
Measure out 100 ml of sample into the measuring cylinder and pour into the porcelain dish. Add a few drops of methyl-orange indicator and titrate with the 0.1M hydrochloric acid until the colour changes from yellow to an orange-pink.

  Alkalinity (as mg/l $CaCO_3$) = ml 0.1M HCl × 50

Permanganate value (4-hour PV)

The measurement of the amount of organic matter in a sample, whether it is of sewage effluent, farm waste, or river water is very important because excessive amounts of organic matter can affect the oxygen balance of a river and threaten aquatic life.

There are two ways of measuring the organic matter, either by using a biological technique as described in the next test, or chemically, by measuring how much of an oxidizing agent is used up (reduced) by the organic matter. In this test, the oxidizing agent is acidic potassium permanganate and the test is carried out at 27°C over four hours in the dark. The results for the samples are compared with a 'blank' which is distilled water. At the end of the test period, the amount of potassium permanganate left (which has not been reduced by the organic matter) is measured by adding iodide solution. The iodide is oxidized by the permanganate to iodine and the amount of this is measured by titrating with sodium thiosulphate solution. The chemistry involved in these reactions is shown at the end of the test.

Apparatus required:
    250 ml wide-necked glass-stoppered bottles
    10 ml measuring cylinder
    100 ml measuring cylinder
    50 ml pipette
    50 ml burette
    Incubator or water bath set at 27°C. (This is the 'standard' temperature for the test as described in water analysis books. However, if this thermostatic equipment is not available, the test can be carried out in a cupboard in the laboratory. The important point to note is that, if results are to be compared from a number of surveys, the temperature of the test should always be about the same.)

Reagents:
    Potassium permanganate solution (0.0025 M). Dissolve 4.0 g $KMnO_4$ in 1 litre of hot distilled water in a large beaker. Cover the beaker and heat it for 2–3 hours at 90–95°C. Dilute to 10 litres with distilled water and leave for several days in a dark cupboard to allow sediment to settle. Carefully pour the supernatant liquid into a storage container.
    Sodium thiosulphate solution (0.25 M). Dissolve 62.05 g $Na_2S_2O_3.5H_2O$ in distilled water and dilute to 1 litre. This solution should be preserved from deterioration by adding 1 ml of chloroform.
    Sodium thiosulphate solution (0.025 M). This is the 'working' solution and is prepared by diluting 100 ml of the 0.25M solution to 1 litre with distilled water. The solution has to be standardized as shown below.
    Potassium iodate solution (0.0042 M). This is used to standardize the

thiosulphate solution. Dissolve 0.892 g of KIO_3 (previously dried at 105°C) in distilled water and make up to 1 litre.

Potassium iodide solution. Dissolve 10 g of KI in 100 ml of distilled water and store in a dark bottle.

Dilute sulphuric acid solution. **Dangerous:** The preparation of this reagent should be carried out in a sink with cold running water. **The acid must be added to the water.** Carefully add 250 ml of concentrated sulphuric acid to approximately 750 ml of distilled water in a 2 litre Pyrex conical flask. During this addition, heat is given out so the flask is cooled by running cold water over the outside of the flask. Transfer the solution to a labelled bottle. Alternatively, a pre-prepared solution can be used.

Starch solution. Mix about 1 g of soluble starch with a small amount of water in a beaker. Add this to about 200 ml of boiling water and stir. Allow to cool with constant stirring and then transfer to a labelled bottle.

Procedure for standardizing the 0.025 M thiosulphate solution

Into a 250 ml conical flask put 5 ml of KI solution, 10 ml of dilute sulphuric acid, 25 ml of KIO_3 solution (this must be measured accurately with a pipette) and about 100 ml of distilled water. A brown colour is produced by the iodine which is oxidized by the iodate. Titrate with the thiosulphate solution until the solution turns a pale yellow colour. Add about 1 ml of starch solution and the solution will turn deep blue. Continue the titration until the blue colour just disappears.

$$\text{Strength of thiosulphate solution} = \frac{0.025 \times 25}{\text{ml of thiosulphate required.}}$$

The strength of the thiosulphate solution should be adjusted so that it is exactly 0.025 M. Alternatively a correction factor should be applied to the calculations.

Procedure for the permanganate value test

Set out a row of bottles appropriate to the number of samples collected and add a further one for the 'blank'. To each bottle, add 50 ml of permanganate solution, 10 ml of dilute sulphuric acid and 100 ml of the sample. To the last bottle, substitute the sample volume with 100 ml of distilled water. The contents of each bottle are mixed by gently swirling; they are then stoppered and placed in the incubator or water bath for four hours.

After this time, the bottles are set out on the bench and to each bottle is added 5 ml of iodide solution. Any remaining permanganate which has not been reduced by the organic matter will oxidize the iodide to iodine and form a brown colour. This is then titrated with thiosulphate using the same procedure as described above for calibrating the thiosulphate solution.

The permanganate value (4-hour PV) = 2 × (ml of thiosulphate used for the blank solution – ml of thiosulphate used for the sample)

If a sample contains a high concentration of organic matter, the permanganate solution may be completely reduced during the incubation period and the solution will be colourless when it is removed after the four hours. In this case the test will have to be repeated using a smaller volume of sample, say 25 ml, and the volume made up with distilled water. The PV value then has to be adjusted: in the example used here, it has to be multiplied by 4 as only 25 ml was used instead of 100.

Chemistry of PV test
In the standardization of thiosulphate:

$$I^- + IO_3^- + 6H^+ \rightarrow I_2 + 6H_2O$$
$$I_2 + 2S_2O_3^- \rightarrow S4O_6^- + 2I^-$$

In the permanganate test:
$$MnO_4^- + 4H^+ + \text{organic matter} \rightarrow MnO_2 + 2H_2O$$
$$MnO_4^- \text{ (remaining)} + I^- + 4H^+ \rightarrow I_2 + MnO_2$$
$$I_2 + 2S_2O_3^- \rightarrow S_4O_6^- + 2I^-$$

Dissolved oxygen test

The measurement of dissolved oxygen (DO) is a very important test for assessing water quality because oxygen is fundamental to life in water. If the oxygen concentration is reduced by polluting substances such as organic matter or reducing agents (for example, sulphide or ferrous ions), then fish and insect life can die or move to cleaner water. The amount of oxygen present in the water is limited by its solubility and this is affected by the water temperature. At normal winter temperatures of 5°C, the equilibrium concentration of oxygen in the water is 12.7 mg/l, but at a summer temperature of 18°C the equilibrium concentration is 9.45 mg/l.

The amount of dissolved oxygen in water is measured either by using a dissolved oxygen meter or else by a titration procedure which is named after Winkler, the Hungarian chemist who discovered it.

The basis of the method is that when concentrated solutions of manganous sulphate and alkali potassium iodide are added to a water sample, a precipitate is formed. This precipitate is initially of manganous hydroxide which is white, but if dissolved oxygen is present then the precipitate immediately turns brown because of the formation of manganic hydroxide. When the sample and its precipitate are treated with strong acid, the precipitate dissolves and the manganic ion oxidizes the iodide present to iodine. This liberated iodine is then titrated with thio-sulphate ion (as described above for the PV test). If the strength of the thiosulphate is exactly 0.0125 M then each ml of titrant used is equivalent to 1 mg/l of dissolved oxygen.

Apparatus required:
    Sample bottles. Ideally these should be of 250 ml capacity, with a sloping neck so that air bubbles don't get trapped, and with a well-fitting plastic or glass stopper.
    3 × 2 ml dropper pipettes
    50 ml burette
    100 ml measuring cylinder
    250 ml conical flask
    10 ml measuring cylinder
    5 ml pipette
    10 ml pipette

Reagents:
    Manganous sulphate solution (50 per cent). Dissolve 500 g $MnSO_4.4H_2O$ in distilled water and make up to 1 litre.
    Alkali-iodide reagent. **Dangerous.** Dissolve 500 g NaOH and 150 g KI in distilled water and make up to 1 litre.
    50 per cent sulphuric acid. **Dangerous.** Put 500 ml of distilled water in a beaker of 2 litre capacity, and place the beaker in a sink filled with cold water. Carefully add 500 ml of concentrated sulphuric acid and stir to mix. Heat is given off by this mixing which is why the beaker is cooled. **Always add the concentrated acid to the water.**

Thiosulphate solution (0.25 M). Dissolve 62.05 g $Na_2S_2O_3.5H_2O$ in distilled water and make up to 1 litre.

Thiosulphate solution (0.0125 M). This is the working solution and is prepared by diluting 50 ml of the 0.25M solution above to 1 litre with distilled water.

Potassium iodate solution (0.0042 M). This is for standardizing the thiosulphate solution. Dissolve 0.892 g potassium iodate, which has previously been dried at 105°C, in distilled water and make up to 1 litre.

Potassium iodide solution. Dissolve 10 g of KI in 100 ml of distilled water and store in a dark bottle.

Starch solution. Mix 1 g of soluble starch in a beaker with a little water. In a separate beaker, boil 200 ml of distilled water and pour the starch/water mix into the boiling water. Stir the solution as it cools and then transfer it to a glass bottle.

Procedure for standardizing the 0.0125M thiosulphate solution

Into a 250 ml conical flask, add 5 ml KI solution, 2 ml of 50 per cent sulphuric acid, 10 ml of KIO_3 solution (measured with a pipette) and about 100 ml of distilled water. A brown colour appears as the iodate oxidizes the iodide to iodine in the acid conditions. Titrate the liberated iodine with the thiosulphate until it becomes pale yellow, then add 1 ml of starch solution. This immediately turns the solution blue. Continue the titration until the blue colour just disappears. Ignore the reappearance of the blue colour on standing.

$$\text{Strength of the thiosulphate solution} = \frac{0.0125 \times 20}{\text{ml of thiosulphate used}}$$

Sample collection

When collecting a sample for the measurement of DO, it is important not to trap air bubbles in the bottle as this will give rise to a false high reading. The following procedure should be followed.

At the sample site, make sure the sample is taken in the main flow of the river or effluent. Gently fill the bottle by inclining it at the water surface (Figure 41), then make sure it is brimful by pushing the bottle below the surface. Tap the sides of the bottle with the stopper to bring any bubbles adhering to the inside to the surface.

Add about 2 ml of the manganous sulphate solution followed by about

Figure 41. Sampling of river water for measurement of dissolved oxygen

2 ml of the alkali-iodide solution. Place the stopper on the bottle firmly and mop up any overflow water with a tissue. This could be alkaline so take care.

A precipitate will have formed in the bottom of the bottle and this now has to be mixed throughout the sample so as to react with all the DO. Holding the bottle with a finger on the lid, shake the bottle vigorously. The precipitate now contains all the dissolved oxygen in the sample and will slowly settle to the bottom of the bottle. This precipitate is stable for a number of days so the laboratory analysis does not need to start immediately.

Analysis procedure:

In the laboratory, remove the stoppers from the samples and carefully add about 4 ml of the 50 per cent sulphuric acid into each bottle using a dropper pipette with the tip below the water surface. The stoppers are then replaced and the excess liquid expelled is mopped up with a tissue. The contents of the bottles are thoroughly mixed by shaking them. The precipitate has now dissolved and iodine has been released to give a clear reddish brown liquid. (At this stage you can assess the DO content because the amount of iodine released, and therefore the intensity of the colour, is directly proportional to the amount of DO. You might consider photographing a row of bottles collected on a survey to illustrate the variation in oxygen content at different sample points.)

For each sample, measure 100 ml from the sample bottle into a conical flask and titrate with the 0.0125 M thiosulphate solution. When the colour has reached a pale yellow, add about 2 ml of starch solution and a deep blue colour is formed. Continue the titration until this just disappears and note down the burette reading. Ignore any blue colour that develops in the flask on standing.

$$\text{Concentration of dissolved oxygen (mg/l)} = \text{ml of 0.0125 M thiosulphate used.}$$

Chemistry of the DO test

$$MnSO_4 + 2NaOH = Mn(OH)_2 \text{ (white precipitate)} + Na_2SO_4$$
$$2Mn(OH)_2 + O_2 = 2MnO(OH)_2 \text{ (brown precipitate)}$$
$$MnO(OH)_2 + 2H_2SO_4 = Mn(SO_4)_2 + 3H_2O$$
$$Mn(SO_4)_2 + 2KI = MnSO_4 + K_2SO_4 + I_2$$

The iodine released is directly proportional to the amount of dissolved oxygen and is titrated with thiosulphate solution according to the chemistry described for the DO test.

Percentage saturation of dissolved oxygen

As mentioned on page 120, the amount of oxygen that can dissolve in water is dependent mainly on the temperature but also to a minor extent on the atmospheric pressure and the salt content. In fresh water, the variations in the atmospheric pressure have a negligible effect whilst the concentration of dissolved salts hardly alters the DO content. However, the temperature effect must be considered. This is especially important if you are comparing results on different days when the water temperature may vary by a couple of degrees Celsius, or where a heated discharge may be warming the river up from one sampling point to the next.

For this reason, DO results are calculated as the percentage saturation (% sat.). At a particular temperature, there is a concentration of DO in water which is in equilibrium with the air above it. The DO results obtained in the sample are compared against this equilibrium value and expressed as a percentage of it. The equilibrium concentrations for different temperature at normal atmospheric pressure are shown in the Appendix.

As an example, suppose a sample is collected from a river which has a temperature of 14°C, and the titration shows that the DO is 8.6 mg/l. However, at this temperature, from the Appendix, the equilibrium concentration of DO is 10.30 mg/l.

$$\% \text{ saturation} = \frac{8.6 \times 100}{10.30} = 83.5\%$$

It is possible that you may collect a sample and the DO content is greater than the equilibrium value. This is called super-saturation and is usually caused by the photosynthesis of aquatic weeds or algae in summer months. These green plants use up the carbon dioxide and release oxygen. The oxygen sometimes is released into the water by the algae or plants at a faster rate than it can diffuse into the air so the water becomes super-saturated.

In Chapter 3, the eutrophication of Strathclyde Park Loch was described when excessive amounts of algae are formed in the loch in the summer months. This results in super-saturation of the loch water, as shown in Table 24 and compared with winter values.

Table 24. Percentage saturation of dissolved oxygen in Strathclyde Park Loch

Date of sample	% sat.
May 1993	200
November 1993	71
May 1994	170
December 1994	75

Biochemical oxygen demand (BOD)

The measurement of biodegradable organic matter is of fundamental importance for controlling pollution of water courses. When organic waste is discharged into water, naturally occurring bacteria use it as food and break it down. This process is called biodegradation and as it proceeds, dissolved oxygen is used up as the source of energy by the bacteria. The greater the amount of organic matter, the more DO is used up so that, if there is a large amount, the DO levels can get dangerously low for aquatic life. Because of this, a measure of the amount of oxygen that an organic discharge will use up when it enters a river is important and it forms the basis of the BOD test.

In the BOD test, the concentration of DO in the water sample is measured firstly at the time of sampling, then five days later after incubation at 20°C in the dark.

As we saw above for the DO test, the amount of oxygen dissolved in water is limited and the BOD of an effluent can exceed the amount available. In these cases, the sample is diluted with a known amount of clean water which has a negligible BOD, to ensure that about half the DO is left after five days incubation. For example, if a sample of sewage effluent is collected which has a BOD of 20 mg/l, the sample is diluted, one part of effluent to four parts of clean water. This may give an initial DO of 8.5 mg/l and, after five days, this will decrease to 3.5 mg/l. The difference between the two readings (5 mg/l) when multiplied by the dilution factor of 4, gives a final BOD of 20 mg/l. The choice of the amount of dilution which will give 50 per cent depletion of the DO over five days is largely a matter of experience and for that reason it is advisable to make a series of dilutions for a particular discharge. In the example given above, it would be best to use two dilutions – 1/4 and 1/10 – to ensure a result. The BOD values of some typical samples are seen in Table 25.

Table 25. BOD values of some typical samples

Sample type	BOD (mg/l)
Clean river water	3
Polluted river water	10
Treated sewage effluent	15
Untreated sewage	300
Cattle slurry	15,000
Milk waste	140,000

Apparatus required for BOD test:
   All glassware used for the DO test
   An incubator which is thermostatically controlled to 20°C. (If this is not available, then a cupboard in the laboratory could be used which has minimal temperature fluctuation, i.e. it does not receive sunlight at any time in the day and is not situated over heating pipes)
   Clean container for dilution water
   1 litre measuring cylinder

Reagents required:
   All the reagents listed for the DO test
   Distilled or deionized water
   Dilution water reagents (this water has to be free of interfering organic matter but has to contain essential elements for the bacteria so they can degrade the organic matter in the sample):

(1) *Phosphate buffer.* Dissolve 3.4 g potassium dihydrogen phosphate (KH_2PO_4) in 50 ml of distilled water. Add 17.5 ml of 1 M NaOH so that the solution should have a pH of 7.2. Add 0.15 g ammonium sulphate ($(NH_4)_2SO_4$, and dilute to 100 ml.

(2) *Magnesium sulphate.* Dissolve 1.0 g magnesium sulphate ($MgSO_4.7H_2O$) in 100 ml of distilled water.

(3) *Calcium chloride.* Dissolve 1.1 g of anhydrous $CaCl_2$ in 100 ml of distilled water.

(4) *Ferric chloride.* Dissolve 0.025 g of $FeCl_3.6H_2O$ in 100 ml of distilled water.

These nutrient solutions are added at the following proportions for each litre of distilled water:

(1) 1.25 ml   (2) 2.5 ml   (3) 2.5 ml   (4) 0.5 ml

The water should be aerated before use to ensure that it is saturated with dissolved oxygen, and should be used as soon as possible. It is recommended to discard it after one week.

Procedure for BOD test
The analysis of BOD ideally should be carried out on the day the samples are taken. If this is not possible, the samples should be stored in a 'fridge overnight. The next day, the samples should be allowed to warm up to room temperature before performing the analysis; this should take about an hour.

The required volume of sample is poured into a 1 litre measuring cylinder and diluted to the required proportion with the dilution water. For example a 1/10 dilution will require 100ml of sample and 900 ml of dilution water. The diluted sample is mixed and then poured into two clean 250 ml glass bottles (as described for the DO samples) so that they overflow. Ensure that there are no air bubbles adhering to the insides of the bottles by tapping the sides and then insert the stoppers firmly.

The dissolved oxygen concentration in one of the bottles is measured immediately, as described in the procedure on page 123. The other bottle is placed in the incubator or cupboard for five days after which time the DO is tested again. For each batch of samples, the BOD of a 'blank' is also determined by filling two bottles with the dilution water and measuring its DO before and after five days.

$$\text{BOD value in mg/l} = \frac{[(a - b) - z(x - y)] \ (z + 1)}{(z + 1)}$$

a = initial DO in sample
b = final DO in sample
x = initial DO in blank
y = final DO in blank
z = volume of dilution water to one volume of sample

For high dilutions, say 1 part of sample to 49 parts of dilution water, this formula can be modified to:

$$\text{BOD value in mg/l} = [(a - b) - (x - y)] \ (z + 1)$$

pH value

The pH value of a water sample is a measure of whether it is acidic or alkaline. The pH is defined as 'the logarithm of the reciprical of the hydrogen ion concentration expressed as moles/litre'. It has a scale of 0–14, where a value of 7 is neutral. Values less than 7 are acidic whilst those above 7 are alkaline. Pure water has a pH of about 5.6 because carbon dioxide dissolves in it to form a weak solution of carbonic acid. Most river waters are about neutral but, as described below under 'acid rain', values below pH 5.0 can be found.

A particular problem arises in some rivers and lakes in summer time when pH values can be as high as 9.0. This occurs because of the photo-synthesis of water weeds or algae; they utilize the carbon dioxide in the water and create an imbalance in the equilibrium between the carbonates and bicarbonates:

In natural water $H_2O + CO_2 \leftrightarrow H^+ + HCO_3^-$, i.e. slightly acid

When photosynthesis occurs, the CO_2 is used up and disturbs the equilibrium of the carbonates and bicarbonates in the water:

$$2HCO_3^- \leftrightarrow CO_3^{2-} + H_2O + CO_2$$

As the carbon dioxide is used up, this reaction shifts to the right and the carbonate dissociates:

$$CO_3^{2-} + H_2O \rightarrow HCO_3^- + OH^-, \text{ i.e. water becomes alkaline}$$

Returning to the example of Strathclyde Park Loch, in winter months, the pH is slightly alkaline at about 7.5 but during the summer of 1995, when the water became super-saturated with DO because of the photosynthesis of the algae, the pH increased to 9.2.

Procedure
pH is usually measured with a pH meter. Either the meters are fitted with two electrodes and the potential difference between the two is measured when they are put into the test solution, or they have a 'combination' electrode which incorporates both electrodes in the one unit. Before the samples are tested, the meter must first be calibrated by immersing the electrode(s) in a solution of known pH – this is called a buffer solution. There are specially prepared tablets of chemicals which, when dissolved in a set volume of distilled water, produce a solution of known pH. There are operating instructions supplied with each instrument and these should be read carefully before carrying out the test.

Tip drainage or mine-water discharge

Chapters 6 and 7 described the problems caused by the drainage water from tip sites and the release of ferruginous (iron-bearing) mine water from abandoned coal mines. As seen in those sections, the appropriate tests are for organic matter (either as 4-hour PV or BOD), chloride, pH and iron. All these procedures have already been described except for iron.

Iron

When iron enters a water course, either from a tip site or an abandoned mine, it is quite often in its reduced, ferrous Fe^{2+}, state which is soluble. The action of flowing into a stream which has oxygen dissolved in it oxidizes the iron to its insoluble ferric state Fe^{3+}. The oxidized iron is present as fine particles and these adhere to surfaces to form bright orange deposits. The concentration of iron in mine water varies considerably but may be several hundreds of mg/l whilst that for tip site drainage is much less, often tens of mg/l. For normal river waters, the values are usually about 0.5 mg/l, but there are exceptions to this depending on the geology and industry of the area. At concentrations of 2 mg/l, the iron gives a slight orange turbidity to the water.

There are a number of different tests for iron but in this procedure both oxidized forms of iron react with thioglycollic acid in ammonia solution to give a reddish-purple colour. The intensity of the colour is in proportion to the concentration according to Beer's Law. A series of standard solutions are prepared and their colour intensity measured with a spectrophotometer to produce a calibration graph. The colour in the sample is measured and its concentration determined using the calibration graph.

Reagents required:

Ammonium hydroxide solution. In a fume cupboard, dilute 37 ml of concentrated ammonia solution to 100 ml with distilled water.

Citric acid solution. Dissolve 20 g citric acid in 100 ml of distilled water.

Thioglycollic acid. Use in its concentrated form.

Stock standard iron solution. Dissolve 1.404 g of ferrous ammonium sulphate $Fe(NH_4)_2(SO_4)_2.6H_2O$ in 50 ml of distilled water to which has been added 20 ml of concentrated sulphuric acid. **(Dangerous: make sure you add the acid to the water.)** Add 0.02 M potassium permanganate solution drop by drop until a faint pink colour persists. Dilute to 1 litre with distilled water. 1 ml of this solution contains 0.2 mg Fe.

Working standard solution. Dilute 50 ml of the stock standard solution to 1 litre with distilled water. 1 ml of this solution contains 0.01 mg Fe.

Procedure

Pipette 50 ml of sample into a beaker, add 2 ml of the citric acid solution, 0.1 ml of thioglycollic acid and sufficient of the ammonia solution (usually about 2 ml) to make the solution ammoniacal. Mix and allow to stand for five minutes. Prepare a reagent blank by adding the same quantities of reagents to 50 ml of distilled water.

The calibration graph is prepared by setting up a row of 5 × 50 ml volumetric flasks, and adding the following to each flask:

Flask 1 (blank) – 2 ml of citric acid, 0.1 ml of thioglycollic acid, 2 ml of ammonia solution

Flask 2 (1.0 mg/l Fe) – 5ml of working standard solution + same volume of reagents as above

Flask 3 (2.5 mg/l Fe) – 15 ml of working standard + reagents
Flask 4 (5.0 mg/l Fe) – 30 ml of working standard + reagents
Flask 5 (10 mg/l Fe) – 2.5 ml of stock standard solution + reagents

Assess the intensity of the colour of the standard solutions and the samples by measuring out a portion of each coloured solution into a 10 mm spectrophotometer cell and place these in the spectrophotometer which has been set at 535 nm wavelength. Prepare a calibration graph of optical density against concentration and read off the values for the samples.

Acid rain

The environmental issues associated with the acidification of precipitation (rain, snow and hail) have been described in Chapter 8. From this you learned that the acidification is caused largely by the emissions of acid gases (sulphur dioxide and oxides of nitrogen) into the atmosphere from power stations and large industrial complexes. A further significant source is vehicle exhausts. The acidifying gases of SO_2 and NOx react with sunlight and water vapour to produce sulphuric and nitric acids which subsequently fall to earth as 'acid rain'.

The acidity of the rain, or the receiving waters such as streams and lakes, is measured by the pH value and the procedure for this is given in the manufacturer's instructions for the operation of the particular pH meter that you have in your laboratory.

An investigation into the extent of pollution in a particular water body caused by acid rain requires the measurement of pH, sulphate, nitrate and alkalinity. This last parameter is useful because it gives an indication of the calcium content of the water which, in turn, reflects the nature of the geology of the catchment. The procedure for it has already been given. The measurement of nitrate is a difficult and hazardous procedure in the absence of specialist equipment and so has not been described in this section. It is possible that you could have your collected samples analysed for nitrate by your local water laboratory using their instruments. The measuring of the concentration of sulphate described here is a turbidimetric procedure.

Sulphate

The sulphate ions in the water sample are precipitated in acid solution by the addition of barium chloride solution. The insoluble barium sulphate forms fine crystals of uniform size that give a cloudiness to the water (turbidity). The greater the turbidity the more sulphate is present. The turbidity is measured with a spectrophotometer and the amount of sulphate present in the sample is determined by comparing its turbidity with standard solutions.

Reagents required:

Conditioning reagent. Working in a fume cupboard, dissolve 75 g NaCl in 300 ml of distilled water, add 30 ml of concentrated hydrochloric acid **(Dangerous)** followed by 100 ml of 95 per cent isopropyl alcohol. Add 50 ml of glycerol and mix well.

Barium chloride crystals. These should be fine enough to pass through a 20 mesh sieve but be retained by a 30 mesh sieve.

Standard sulphate solution. Dissolve 0.1479 g of anhydrous sodium sulphate, Na_2SO_4 in distilled water and make up to 1 litre in a volumetric flask. 1 ml of this solution contains 0.1 mg SO_4^{2-}.

Procedure

Before the test is started, check that the samples collected are not already turbid with fine suspended solids. If they are, filter them before starting the test and use the filtrate. If the samples are coloured (e.g. peaty) then this must be compensated for in the test.

Measure into a 250 ml conical flask an amount of sample containing no more than 4 mg of sulphate. (As a guide, rainwater contains about 1–10 mg/l SO_4^{2-} whilst river water can contain 5–500 mg/l.) Make the volume up if necessary to 100 ml with distilled or deionized water. Add 5 ml of the conditioning reagent and mix. The solution is now stirred constantly and at a consistent speed for the rest of the procedure. This is to ensure that the barium sulphate crystals are evenly sized. The stirring is done either manually using a glass rod or with a magnetic stirrer: this should be set so that it stirs rapidly but doesn't splash.

While stirring, add about 0.5 g barium chloride crystals and continue to stir for a further one minute exactly. Pour the turbid solution into 40 mm spectrophotometer cell and fill another 40 mm cell with distilled water. (If the samples collected are coloured then the 'blank' for measuring the turbidity against should be one of the samples.) Measure the turbidity at 420 nm and take readings every 30 sec for about 4 min. Note the maximum reading obtained.

Prepare a calibration graph by measuring appropriate amounts of the standard solution into a number of 250 ml conical flasks to cover a range up to 4 mg sulphate ion (i.e. up to 40 ml of standard solution). Make up the volume in each flask to 100 ml with distilled or deionized water. Carry out the procedure as above. From a measure of the optical density for each standard solution, prepare a calibration graph and calculate the amount of sulphate in the samples. Express the results as mg SO_4^{2-} /litre.

Investigating acid rain

Acid rain affects the water quality of streams only if there is insufficient neutralizing capacity in the soils and rocks in the catchment. The streams most likely to be affected will be those where the catchment geology is of hard igneous rocks such as granite and these can be checked before-hand by looking at a geological map of the country.

It is possible though to check the acidity of rainwater and to see the influence of wind direction on the pH. To carry out such an investigation, a special rainfall collection device has to be prepared to ensure that the collected rainwater is not contaminated.

Rainwater collector
The ideal rainwater collector comprises a large, clean polythene funnel which is fitted over a polythene bottle. The collecting device should be fitted to a post about 1 metre from the ground and in an open area so that the rainwater is not firstly in contact with leaves from a tree. A partic-ular problem that has to be dealt with is that birds may sit on the edge of the funnel and contaminate the rain water with their droppings! The birds can be deterred by putting a circle of chicken wire around the edge of the funnel as shown in Figure 42.

If you do not have a suitably sized polythene funnel, you can make an effective rainwater collector from mineral water bottles. These are very suitable because they were clean before the mineral water was put into them. You'll need two sizes of bottle, say a 5 litre one and a 500 ml one. The top part of the large bottle is cut off from the rest of it to form a funnel when it is inverted, whilst the bottom part of the collector is the smaller bottle which has also been modified to accommodate the funnel, as shown

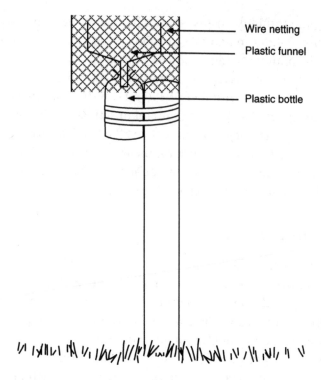

Figure 42. Rainwater collector for acid rain study (1)

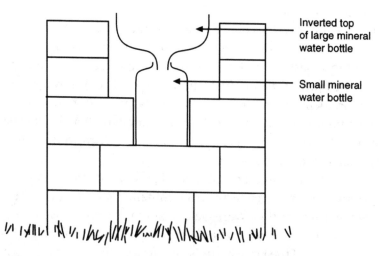

Figure 43. Rainwater collector for acid rain study (2)

in Figure 43. Again, this collector should be placed in an open space above ground level: it could be placed on a pile of bricks.

Rainwater should be collected from the rainwater collectors on each occasion that it rains. You may find that there is enough rainwater to analyse for both pH and sulphate, but if the amount is limited then just carry out a measurement of pH. On each occasion that you collect rain, make a note of which direction the weather is coming from. One way to find this out is to look at the weather map published in a daily paper; those in *The Daily Telegraph* or *The Guardian* are particularly good.

As you carry out the project over a period of time you will be able to compile a table of rainfall pH and sulphate content, and the wind direction. These data can be represented as a 'wind rose', as shown in Figure 44 for some rainfall collected in south-west Scotland.

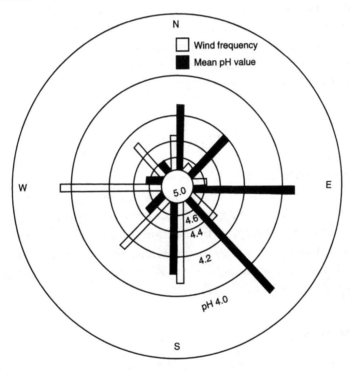

Figure 44. Effect of wind direction on rainwater pH in Scotland

Snow

On some occasions the precipitation will occur as snow. The snow is likely to accumulate in the funnel of the rain collector but the whole device can be brought indoors to allow the snow to melt and flow into the collecting bottle. You could also collect snow from an area which has not been disturbed. Simply go to the selected area and scoop the snow into a beaker that has previously been cleaned and rinsed with distilled or deionized water. As with rainwater, measure the pH and sulphate content and look for variations according to wind direction.

In some parts of the world where snowfall is frequent such as in Canada and Scandinavia, snow may be discoloured by pollutants that the flakes have collected during their formation. This discoloured snow is usually acidic because the pollutants causing the discoloration are soot particles from chimneys which have also emitted sulphur dioxide and nitrogen oxides.

You can check the snow sample you have collected for soot particles by first melting it in the beaker and then filtering it through a filter paper. The filter paper is then dried and examined for the presence of black particles. You can compare the cleanliness of snow which you have collected from the same area on different occasions and again see whether the wind direction affects the quality of the snow.

Air pollution

Throughout this book there have been many references to air pollution and the influence that traffic has on poor air quality, particularly in towns and cities. There are a number of simple experiments that can be carried out to monitor air quality. In this section, procedures are described for measuring particulates, acid gases and exhaust emissions whilst the effects of air pollution can be measured by looking for yeast cells on leaves or by counting the variety of lichens.

Particulates

Traffic exhausts, especially from diesel engines, contain large numbers of fine particles and these can be hazardous to health because they can be breathed directly into the lungs where they aggravate the lung tissue. Their presence can be measured either in the air in the street (the ambient air) or close to a vehicle.

Procedure for measuring particulates in ambient air

This experiment should be carried out on a dry, calm day. Select your measuring site which can be on a window ledge or close by a street. You may need an electrical extension lead if your monitoring site is well away from a power point. Plug in a small vacuum pump – a suitable pump will be an aquarium pump but using the suction end – and connect it to a filter funnel onto which you have placed a circle of dampened filter paper. Switch on the pump and let it run for a predetermined period of time, say one hour. Disconnect the pump and examine the filter paper for accumulated particles. If this procedure is followed in a number of different locations, such as indoors and outdoors, or town and country areas, you can compare the intensity of the colour of the filter papers as a measure of air quality as long as you use the same period of time for suction.

Procedure for measuring particulates in a vehicle exhaust

Another experiment to carry out is to check on the emissions from individual vehicles. You will be aware of government plans to improve air quality by persuading us that we should use our cars less. One problem in towns and cities is that there is so much traffic, that the vehicles spend a lot of time stationary with their engines running. This produces polluting gases and particulates.

In this procedure, you will be able to assess the amount of particulates that are emitted from a stationary vehicle. You will need:

An electrical extension cable
A vacuum cleaner (a cylinder type or one with a hose attachment)
A piece of filter paper large enough to be fitted over the end of the suction hose and secured with an elastic band

The vehicle to be tested is parked in a convenient position in the open air (do not do this experiment in a confined space like a garage) close to where the vacuum cleaner is connected to the extension cable. Have the owner start the vehicle and let it idle in neutral gear. Switch on the vacuum cleaner and, standing beside the vehicle, direct the hose with its attached piece of filter paper so that it is about half a metre beyond the end of the exhaust pipe. Suck in the exhaust gases for about 10 minutes, have the engine switched off and then disconnect the vacuum cleaner. Examine the colour of the filter paper and compare it with another piece that has been used to suck air without the vehicle present.

Using this technique, you can compare the results for different vehicles or else the same vehicle before and after the engine has warmed up. You are probably aware that, on cold winter days, a vehicle may have reached its destination (if it's a short journey) before the engine has warmed up and so the pollution levels are greater in winter.

Biological effects of air pollution

Lichens as a measure of air pollution
We learned in Chapter 11 about the value of indicator species for monitoring pollution. Just as invertebrate organisms are an important group of indicator organisms for measuring water quality, a useful indicator for air quality is the variety of lichens that are present. Lichens are a group of very slow-growing organisms that you find as encrustations on stonework or as a flaky growth on tree barks. They are unusual organisms because they consist of fungal threads and microscopic green algae which live together and function as one organism. The term for this relationship is mutualism – a symbiotic relationship that is beneficial to both participating species. There are a number of different species and Cover Illustration 12 shows an example. They have different sensitivities to pollution so the cleaner the air, the more species are present.

Figure 45 shows the results of a survey of lichens at increasing distance from a city centre, whilst Figure 46 gives an indication of their distribution in England and Wales.

The reason why lichens are sensitive to air pollution is because they

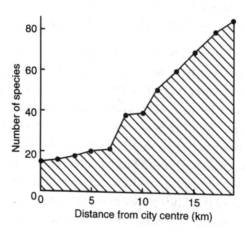

Figure 45. Lichen numbers at increasing distance from a city centre

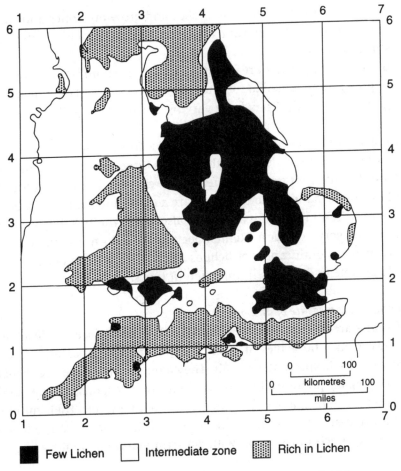

Few Lichen | Intermediate zone | Rich in Lichen

Figure 46. Lichen distribution over England and Wales

do not have roots and rely instead on the absorption of rainfall, and the nutrients in it, for growth. Their surfaces lack the protective layer that the leaves of plants have which allows them to block out pollutants. As a result the pollutants can accumulate within the lichens and reach levels where they break down the chlorophyll molecules responsible for photosynthesis.

Procedure for monitoring air pollution using lichens
There are three main groups of lichens:

Shrubby type. These look like leafless bushes. They are grey/green in colour, about 2–5 cm high, and are very sensitive to pollution.

Leafy type. These have flat leafy lobes and grow on stones and trees. They also are grey/green and are more tolerant of pollution than the shrubby type.

Crustaceous type. These are found in a variety of colours – white, orange or green – and look like splashes of paint on surfaces on which they grow very tightly. They have varying sensitivity to pollution.

Using an Ordnance Survey map, move out from a source of pollution such as a city centre or a factory which emits fumes into the atmosphere. Look out for different types of lichen but always use the same substrate, i.e. old walls or mature tree bark. Prepare a map showing the number of species of lichen in different areas and also note whether they are rare, common or very common. Remember that you do not have to be an expert in the identification of lichens but with a little experience you should be able to distinguish the different types.

Leaf yeasts as a measure of air pollution[1]

Leaf yeasts are microscopic fungi that live on the surfaces of deciduous tree leaves and the leaves of grasses, shrubs and herbs. They are very sensitive to toxins present in the air. They have an advantage over lichens for studying air pollution because of their faster response. Lichens take years to grow whilst leaf yeasts develop in about a week. Their numbers vary considerably according to the air quality and enable you to be able to map the effects of pollution with a discrimination of about 1 km^2.

Procedure for mapping air pollution using leaf yeasts

In this procedure the leaves of trees from different places in your study area are collected. In the laboratory, the yeast cells on the leaves are allowed to fall onto a special nutritious medium which enable them to grow into small visible colonies. The colonies are counted and their numbers related to sources of air pollution.

First draw a map of the area you wish to study or else use an Ordnance Survey map: the series 1:25,000 is particularly useful for this study. Your study area should be based on 1 km square grids.

For each of the squares on the map, visit the area and look for suitable trees. Ash, sycamore and limes are very common in towns. Select between three and five trees of the species you have decided to sample in each 1 km square, one in the middle and the others nearer the edge. Choose trees where you can collect leaves by reaching up from the ground. Record their location on the map and note particularly whether they are beside a busy or quiet road, or in a field or park.

Sample the trees on the designated day in spring or summer. Try to avoid autumn because the number of yeasts changes as autumn advances. From each tree collect enough leaves for you to cut out nine discs – about three leaves should do. Take leaves from different sides of the tree and don't collect from newly opened buds because the yeasts will not have developed on them. If you have sampled four trees in the 1 km square, you should have about twelve leaves.

When you have collected leaves from all the individual 1 km squares, return to the laboratory and prepare a dust-free area on a table or laboratory bench. Wash the area with hot water and detergent and then wipe it with a paper towel which has been dipped into 70 per cent alcohol.

Cut four to five discs from each leaf using a 1 cm cutter and place onto a clean piece of paper which has been labelled with the location, one piece of paper for each tree sampled. The discs are placed on the paper so that the lower surface is uppermost. There should now be about 50 discs for each 1 km square.

For each grid square, have five petri dishes (with their lids) prepared containing 1.5 per cent malt extract agar in them. Label each dish with a letter and assign a number for each grid square, i.e. you'll have 1A, 1B, 1C, 1D, 1E; 2A, 2B, etc. Place each dish upside down on the cleaned bench, lift the base containing the medium off the lid and place it upside down next to the lid. Do not turn the dish over or touch the medium otherwise it may pick up yeasts from the air or from your fingers.

On the inside of each lid put nine small blobs of Vaseline using a sterile piece of glass rod or similar implement. Using a pair of forceps, place the nine discs from each individual tree onto the lid so that their smooth upper sides are stuck to the Vaseline. Replace the lid over the base containing the agar. Repeat for the other trees in the 1 km square. Pile the Petri dishes for each 1 km square on top of each other and secure them together with tape.

Invert the pile so that the leaves are uppermost and are suspended over the agar. The leaf yeasts will fall off the leaves and land on the agar where they will grow.

Place the individual piles of dishes onto a tray and put them in an area where they will not receive any direct sunlight. After 24 hours, invert the piles of dishes so that no more yeasts will drop off the leaf discs.

Leave the stacks for a further 48 to 72 hours (it doesn't matter what time interval you choose as long as you use the same interval for each test). Examine the agar medium and you will find pink colonies of yeasts have developed from each individual yeast that has fallen off the disc. Ignore white glistening colonies or fluffy white or grey ones as these are caused by other organisms.

Using a magnifying glass, count the number of each pink colony under the individual discs and record the results. If the colonies have coalesced because they have been incubated for too long, then reject the result for that disc. If the number of colonies under a disc is >50 then estimate the number.

From all the individual dishes for each 1 km square, record the median value and classify the results according to this scale:

$$30+ \quad = \text{very good}$$
$$14–29 = \text{good}$$
$$6–13 \quad = \text{moderate}$$
$$2–5 \quad = \text{poor}$$
$$0–1 \quad = \text{very poor}$$

Transfer the readings you have obtained to the map of your sample area and try and relate the results to sources of pollution. Remember to take into account the wind direction on the days before the sampling took place.

Note

1. P. Dowding and D.H.S. Richardson, 'Leaf yeasts as indicators of air quality in Europe', *Environmental Pollution*, 66, 1990, 223–35.

Postscript

This book has described a range of environmental problems which are encountered in the UK and in other developed countries at the close of the twentieth century. We are fortunate that we are living in a time when our environment in the UK has improved greatly even though there are more of us and we are emitting more waste, whether it be fumes, effluents or rubbish. This is because there are many more controls and limits on how much we can discharge into the environment. There are always new pressures though: at the time of writing at the start of 1998, there is concern about air quality from the rapidly increasing numbers of private cars in towns and cities, and about groups of chemicals that can alter the endocrine systems responsible for reproduction of aquatic life. Some of these chemicals have produced male characteristics in marine molluscs whilst others have resulted in the feminization of male fish in some rivers.[1]

These environmental pressures are being brought under control by new environmental laws. The European Union plays a major part now in environmental legislation and most environmental improvements in the UK are as a result of laws emanating from Brussels rather than London. In the water sector, there are EC Directives covering the quality of drinking water and bathing water, there are defined quality levels for waters for fish life and shellfish growing, there are limits on discharges for a great many hazardous substances, and there are standards for the quality of effluents discharged into rivers and the sea. Work is in progress on defining 'good ecological quality' waters with the aim that member states of the EU will achieve this status by 2010. There are also EU laws covering solid waste disposal, the amount of packaging used and its recycling, and there are targets for improving air quality and reducing emissions of acid gases and CO_2, limits on noise emitted from a wide range of machinery, controls on the movement of radioactive and other hazardous wastes between countries, and many more environmental pollutants are covered by various Directives.

Referring back to Chapter 1 and the growth of the human population, the environmental issues we have been concerned about in this book pale into insignificance when compared with the problems encountered in those countries of the developing world where poverty, overpopulation and lack of food, water and education threaten the lives of people living there.[2] In these countries, the availability of clean drinking water is probably one of the most important factors that affects people's day-to-day living. Some stark statistics convey the extent of the problem:

World-wide: 27 countries are short of water
a quarter of the population has no safe water to drink
nearly a half have no proper sanitation
4 million children die each year of waterborne diseases.[3]

In our advanced nations, we are demanding higher and higher standards for our water supply which involves the expenditure of millions of pounds on new treatment processes. In the developing world, the concern is about just getting water. Growing populations and depleted supplies involve millions of people (usually women) in long walks to a muddy waterhole to fill up a container of water for the family's needs. If only such human effort could be channelled into more productive work and education, then there would be some hope for their future. Some advances in the provision of drinking water in the developing world have been made, particularly by the efforts of charities such as WaterAid, Christian Aid, Save the Children, etc. which are involved in thousands of projects world-wide, such as drilling new boreholes to underground supplies, providing simple pumping equipment, pipes and storage tanks and setting up effective sanitation systems. However, these are local efforts which are alleviating local problems. Of more concern is the way some countries are exploiting water supplies for their own people and denying them to neighbouring countries. This happens particularly in areas of scarce supply where river water is being extracted in one country and this does not leave enough for the neighbouring country downstream. There are special problems in the Middle East with Jordan, Israel, Palestine, Turkey, Iraq, Iran and Syria all trying to secure the diminishing water supplies for themselves. The last Secretary General of the United Nations, Boutros Boutros-Ghali, said, 'The next war in the Middle East will be fought over water, not politics.'

Notes

1. T. Colburn, J.P. Myers and D. Dumanoski, *Our Stolen Future*, Little, Brown and Company, Boston, 1996.
2. K.T. Pickering and L.A. Owen, *An Introduction to Global Environmental Issues*, Routledge, London, 1994.
3. S. Postel, *The Last Oasis – Facing Water Scarcity*, Earthscan Publications Ltd, London, 1992.

Appendix

Concentration of dissolved oxygen in water corresponding to 100 per cent saturation at different temperatures and at 760 mmHg atmospheric pressure

Water temperature (°C)	Concentration of dissolved oxygen (mg/l)	Water temperature (°C)	Concentration of dissolved oxygen (mg/l)
1	14.2	14	10.30
2	13.8	15	10.10
3	13.4	16	9.85
4	13.1	17	9.65
5	12.7	18	9.45
6	12.4	19	9.30
7	12.1	20	9.05
8	11.8	21	8.90
9	11.6	22	8.73
10	11.3	23	8.55
11	11.0	24	8.40
12	10.7	25	8.25
13	10.5		

Example

If a river water sample contains 8.6 mg/l dissolved oxygen at 14°C, then the percentage saturation of the sample is:

$$\frac{8.6}{10.30} \times 100 = 83.5\%$$

Useful addresses

Since April 1996 in the UK, the management of all sectors of the environment – land, water, air and radioactivity – has been controlled by three organizations: the Environment Agency for England and Wales, the Scottish Environment Protection Agency for Scotland, and the Northern Ireland Environment and Heritage Department in the Department of the Environment, Northern Ireland.

Each of these organizations produces a variety of educational material for use in schools and colleges and can answer specific questions about pollution in their respective countries. Both the EA and SEPA have produced 'State of the Environment Reports' which are particularly useful sources of information on a range of environmental issues. The addresses to contact initially are:

Environment Agency (EA)
Rio House
Waterside Drive
Aztec West
Almondsbury
Bristol BS12 4UD
Tel: 01454 624400

Scottish Environment Protection
    Agency (SEPA)
Erskine Court
The Castle Business Park
Stirling
FK9 4TR
01786 457700

Department of the Environment for Northern Ireland
Environment and Heritage Service
Calvert House
23 Castle Place
Belfast
BT1 1FY
Tel: 01232 254754

The above are the 'regulatory' authorities and they control pollution from various sources. The UK government has its own Department of the Environment, Transport and Regions in London which makes policies about the protection of the environment. Its main office is:

Department of the Environment, Transport and Regions
2 Marsham Street
London
SW1P 3EB

The corresponding organization in Scotland is:

The Scottish Office Agriculture, Environment and Fisheries Department
Victoria Quay
Leith
Edinburgh
EH6 6QQ

The Department of the Environment for Northern Ireland is at:
Northland House
3 Frederick Street,
Belfast
BT1 2NR

In England and Wales, the water industry has been privatized. These public limited companies are responsible for the supply of safe drinking water and for the purification of sewage so that it meets the standards set by the Environment Agency. They can supply details of the chemical and bacteriological quality of drinking water and the standards of sewage treatment works' effluents. They may permit you to visit their treatment works. The companies' addresses are:

Northumbrian Water	Severn Trent Water	South West Water
Regent Centre	2297 Coventry Road	Peninsula House
Gosforth	Birmingham	Rydon Lane
Newcastle-upon-Tyne	B26 3PU	Exeter
NE3 3PX		EX2 7HR
Thames Water	Hyder plc	Wessex Water
14 Cavendish Place	Plas y Ffynnon	Wessex House
London	Cambrian Way	Passage Street
W1M 0NU	Brecon	Bristol
	Powys	BS2 0JQ
	LD3 7HP	
Yorkshire Water	North West Water	Anglian Water
2, The Embankment	Dawson House	Anglian House
Sovereign Street	Great Sankey	Ambury Road
Leeds	Warrington	Huntingdon
LS1 4BG	WA5 3LW	PE18 6NZ

In Scotland, the treatment of sewage and the supply of drinking water are still administered by public authorities. These are:

North of Scotland Water	East of Scotland Water Authority
Authority	Pentland Gait
Cairngorm House	597 Calder Road
Beechwood Park	Edinburgh
Inverness	EH11 4HJ
IV2 3ED	

West of Scotland Water Authority
419 Balmore Road
Glasgow
G22 6NU

For Northern Ireland, the water services are based in the region's Department of the Environment at the address given above.

There are many other organizations which can provide background information on environmental issues. This list will help you :

Environmental Information, a guide to sources
British Library Section
Turpin Distribution Services Ltd
Blackhorse Road
Letchworth
Herts SG6 1HN

UK Directory of Environmental Databases
ECO Environmental Information Trust
10–12 Picton Street
Montpelier
Bristol BS6 5QA

Who's Who in the Environment – separate books for England and Wales
The Environment Council
21 Elizabeth Street
London SW1W 9TR

Who's Who in the Environment (Scotland)
Scottish Environmental Education Council
Department of Environmental Science
University of Stirling
Stirling FK9 4 LA

National Association of Waste Disposal Contractors (NAWDC)
Mountbarrow House
6–20 Elizabeth Street
London SW1W 9RD

Chartered Institution of Water and Environmental Management
15 John Street
London WC1N 2EB

Water Services Association of England and Wales
1 Queen Anne's Gate
London SW1 H 9BT

English Nature
Northminster House
Peterborough
PE1 1UA

Scottish Natural Heritage
12 Hope Terrace
Edinburgh
EH9 2AS

Countryside Council for Wales
(Cygnor Cefn Gwlad Cymru)
Plas Penrhos
Ffordd Penrhos
Bangor
Gwynedd LL57 2LQ

Royal Commission on Environmental Pollution
Church House
Great Smith Street
London
SW1P 3BZ

The Natural Environment Research Council has a number of centres dealing with environmental issues:

British Geological Survey Kingsley Dunham Centre Keyworth Nottingham NG12 5GG	Institute of Freshwater Ecology Windermere Laboratory The Ferry House Far Sawry Ambleside Cumbria LA22 0LP
Institute of Terrestrial Ecology Monks Wood Abbots Ripton Huntingdon PE17 2LS	Plymouth Marine Laboratory Prospect Place West Hoe Plymouth PL1 3DH

There is a great deal of information on environmental issues on the Internet. Rather than give details of the individual sites, the following are well worth 'browsing':
European Environment Agency at http://www.eea.eu.int/

There are reports, articles, brochures and papers as well as a collection of environment-related Internet links.
The England and Wales Environment Agency has a site at:
http://www.environment -agency.gov.uk.

The Scottish Environment Protection Agency's site is at:
http://www.sepa.org.uk

Information on air quality throughout the UK is available from the site run by AEA. This is found at:
http://www.aeat.co.uk/netcen/aqarchive/bkgmaps/index.html

Index

malaria 46
Manchester 13, 93
manganese 30 56
mayfly larvae 82, 107
mercury 18
mesotrophic 29
methane 13, 25, 59, 96
methaemo globinaemia 44
micro-organisms 8
Middle Ages 101
milk solids 15
milt 55
mine water pollution 69
Ministry of Agriculture, Fisheries and Food (MAFF) 45
molybdenum 30
monastery ponds 55
mosquitoes 46, 50
Motherwell 33, 92
mussels 40, 82, 107
mutualism 108

NPK 37
nappy liners 15
National Coal Board (NCB) 70
National Lottery 2
natural gas 84
    substances 14
necessities of life 1
neutralising capacity 78
new towns 27
New York 13
nitrate 16, 17, 23, 24, 26, 39
nitric acid 75
    oxide 75
nitrification 25
nitrogen 7, 30, 59
    dioxide 75, 90
    oxides 89
nitrous oxide 75, 96
Norfolk Broads 30
Norway 40, 54, 75
Nottingham Castle 86
Nottinghamshire 44
nuclear fuel 84
nutrients 7, 10, 17

Ocean Dumping Ban Act 13
oil 15, 16
oilseed rape 42, 49
oligotrophic 29
optical brighteners 17
organic matter 66, 108
    phosphates 16
    phosphorus compounds 25
    substances 15
    sulphur compounds 16, 25
organo-phosphorus insecticide 49, 57
organo- nitrogen insecticide 50
Orkney 40

osmoregulatory mechanism 81
otters 58
overpopulation 144
oxygen 15
ozone 89, 90, 94

PM_{10} 89, 90
Palestine 144
palm oil 16
palm trees 16
palmitic acid 16
palmitin 16
pancreas disease 57
paper bags 14
    mills 18
paralytic shellfish poisoning 40
Paris 93
parr 56
particulate matter 89, 136
percolating filter 9
    sewage treatment works 8
perfume 17
Permanganate Value (4 hr PV) 24, 26
    analysis 117
peroxy acetyl nitrates (PANs) 89
pesticides 45, 108
    manual 45
    spraying equipment 45
pets 50
pH 24, 26
    analysis 128
    meter 76
Philipshill sewage treatment works 27
phosphate 11, 17, 23, 24, 25, 26
phosphines 25
phosphonate 17
phosphorus 30, 55, 59
photochemical smogs 94
photographic developers 18
photosynthesis 24, 67, 98, 105, 125, 128
physico-chemical process 11
phytoplankton 39, 100
pigment 56
pigs 42, 51
plankton 104
plastic 15, 64
    bags 14
    wrapping 15
plating 18
play parks 66
ploughed land 113
Poland 12
polar ice caps 98
pollutants 14
pollution 2
    definition 3
polyelectrolytes 60, 67
polypeptides 16
Portland 93